# Fundamentals of Manufacturing
## Workbook

# Fundamentals of Manufacturing
## Workbook

Philip D. Rufe, CMfgE

Editor

Society of Manufacturing Engineers
Dearborn, Michigan

Copyright © 2006 Society of Manufacturing Engineers

987654

All rights reserved, including those of translation. This book, or parts thereof, may not be reproduced by any means, including photocopying, recording or microfilming, or by any information storage and retrieval system, without permission in writing of the copyright owners.

No liability is assumed by the publisher with respect to use of information contained herein. While every precaution has been taken in the preparation of this book, the publisher assumes no responsibility for errors or omissions. Publication of any data in this book does not constitute a recommendation or endorsement of any patent, proprietary right, or product that may be involved.

Library of Congress Catalog Card Number: 2005936533
International Standard Book Number: 0-87263-845-6

*Additional copies may be obtained by contacting:*
Society of Manufacturing Engineers
Customer Service
One SME Drive, P.O. Box 930
Dearborn, Michigan 48121
1-800-733-4763
www.sme.org

*SME staff who participated in producing this book:*
Rosemary Csizmadia, Production Editor
Frances Kania, Administrative Coordinator
Printed in the United States of America

# Table of Contents

Preface .................................................................................................................. ix
Introduction .......................................................................................................... xi
Mathematical Fundamentals (Questions 1–10) ................................................. 1–6
Physics and Engineering Sciences (Questions 11–30) ..................................... 7–16
Materials (Questions 31–40) .......................................................................... 17–18
Product Design (Questions 41–65) ................................................................ 19–26
Manufacturing Processes (Questions 66–95) ................................................ 27–32
Production Systems (Questions 96–135) ....................................................... 33–40
Automated Systems and Control (Questions 136–145) ................................. 41–44
Quality (Questions 146–170) ......................................................................... 45–50
Manufacturing Management (Questions 171–190) ....................................... 51–56
Personal/Professional Effectiveness (Questions 191–200) ............................ 57–58
Solutions ....................................................................................................... 59–101
    Mathematical Fundamentals ............................................................... 59
    Physics and Engineering Sciences ....................................................... 63
    Materials ................................................................................................ 71
    Product Design ..................................................................................... 73
    Manufacturing Processes .................................................................... 79
    Production Systems .............................................................................. 85
    Automated Systems and Control ......................................................... 91
    Quality .................................................................................................. 93
    Manufacturing Management ............................................................... 97
    Personal/Professional Effectiveness .................................................. 101
References ........................................................................................................ 102

# Acknowledgments

**AUTHOR**

Philip D. Rufe, CMfgE

**TECHNICAL REVIEWERS**

Jack Day, CEM, CMfgE
Michael Flaman, CMfgE
William D. Karr, CMfgT

# Preface

This Workbook is designed to be used in conjunction with *Fundamentals of Manufacturing*, Second Edition. It provides structured practice questions for individuals preparing to take the Manufacturing Technologist (CMfgT) certification examination.

While the objective of this Workbook is to help prepare manufacturing professionals for the Certified Manufacturing Technologist exam, individuals preparing for the Certified Manufacturing Engineer exam may find it beneficial in their preparation process.

# Introduction

## MANUFACTURING CERTIFICATION

Manufacturing is concerned with energy, materials, tools, equipment, and products. Excluding services and raw materials in their natural state, most of the remaining gross national product is a direct result of manufacturing.

Modern manufacturing activities have become exceedingly complex because of rapidly increasing technology and expanded environmental involvement. This, coupled with increasing social, political, and economic pressures, has caused successful firms to strive for high-quality manufacturing engineers and managers.

The Society of Manufacturing Engineers has acted as a certifying body since 1971. The principal advantage of certification is that it shows the ability to meet a certain set of standards related to the many aspects of manufacturing. These standards pertain to the minimum academic requirements needed, but more importantly, they pertain to the practical experience required of a manufacturing engineer or manager.

Many persons currently employed in industry can successfully measure themselves against these standards, but they cannot provide documentation concerning their ability. The certification program is designed to provide successful candidates with documentary evidence of their abilities. The designations Certified Manufacturing Engineer (CMfgE), Certified Manufacturing Technologist (CMfgT), and/or Certified Enterprise Integrator (CEI) are bestowed upon successful candidates.

Philosophically, the purpose of manufacturing certification is to gain increased acceptance of manufacturing engineering and management as a profession and to ultimately improve overall manufacturing effectiveness and productivity.

## PURPOSE AND OVERVIEW

The purpose of this Workbook, in conjunction with the *Fundamentals of Manufacturing*, Second Edition, is to provide structured practice questions for the manufacturing professional preparing to take the Certified Manufacturing Technologist (CMfgT) examination. The workbook can also serve, to a limited extent, as useful practice for individuals preparing for the Certified Manufacturing Engineer (CMfgE) examination.

The major areas of manufacturing reviewed in the Workbook include mathematics, physics and engineering science, materials, product design, manufacturing processes, production systems, automation and control, quality, manufacturing management, and personal effectiveness.

## EXAMINATION SPECIFICS

The Certified Manufacturing Technologist examination is a three-hour, open-book exam consisting of 130 multiple-choice questions. Each major area and its relative emphasis in the exam are listed as follows:

| | |
|---|---|
| Mathematics | 2.1% |
| Physics and Engineering Science | 9.1% |
| Materials | 5.1% |
| Product Design | 13.4% |
| Manufacturing Processes | 14.1% |
| Production Systems | 20.9% |
| Automated Systems and Control | 5.3% |
| Quality | 13.0% |
| Manufacturing Management | 10.8% |
| Personal/Professional Effectiveness | 6.2% |

## ADDITIONAL INFORMATION

Additional study resources for the Certified Manufacturing Technologist exam include, but are not limited to, a pencil and paper practice exam, the web-based or CD-ROM self-assessment program, and *Fundamentals of Manufacturing*, 2nd Edition. The self-assessment program (based on the same topics as the exam) will help to determine a candidate's strengths and weaknesses. Built-in bibliographic references suggest additional study materials.

For more information regarding the exam or additional resources, please contact the Society of Manufacturing Engineers by calling 313-271-1500 or email: training@sme.org. Information also can be obtained on SME's website www.sme.org/certification.

Any questions or comments regarding this Workbook are welcome and appreciated. Please direct questions and/or comments to training@sme.org.

# Mathematical Fundamentals

1. Solve the following equation for $x$.

    $\log_{10} x = 4 - \log_{10}(4x - 2)$

2. Solve the following equations for $x$ and $y$.

    $6x + 3y = 10$
    $3x + 4y = 20$

a) 16.06, 14.06
b) 50.25, −49.75
c) 25.75, 15.75
d) 50.25, 49.75

a) $x = 8, y = 6$
b) $x = -1.33, y = 6$
c) $x = 3, y = 0$
d) $x = 1.33, y = 6$

3. The volume of a cylindrical tank is 10 m³. Assuming the tank's height is twice its diameter, find the tank's diameter.

a) 0.55 m
b) 0.93 m
c) 1.85 m
d) 3.70 m

4. Which of the following lines is parallel to the line with the equation $y = \frac{1}{2}x + 4$?

a) $y = -\frac{1}{2}x + 4$
b) $y = 2x + 4$
c) $x = \frac{1}{2}y + 4$
d) $2x = 4y - 4$

5. Find the distance between the following two points.

   $x_1 = 10$, $y_1 = 10$
   $x_2 = 2$, $y_2 = 3$

   a) 6.6
   b) 8.4
   c) 10.6
   d) 12.4

6. Company XYZ wants to randomly select a committee of three people out of the five-person engineering section. Diane is one of the five people. What is the probability that Diane will be on the committee?

   a) 20%
   b) 50%
   c) 75%
   d) 60%

## Mathematical Fundamentals

7. What is the average height and standard deviation of adult American men based on the following sample?

| Person | Height (in.) |
|--------|--------------|
| 1  | 61 |
| 2  | 72 |
| 3  | 65 |
| 4  | 78 |
| 5  | 72 |
| 6  | 68 |
| 7  | 70 |
| 8  | 70 |
| 9  | 64 |
| 10 | 80 |

a) 70 in., 5.6 in.
b) 68 in., 5.9 in.
c) 70 in., 5.9 in.
d) 68 in., 5.6 in.

8. A rolling operation produces parts that are normally distributed with a mean thickness of 0.220 in. and a standard deviation of 0.002 in. If the specification calls for a thickness of 0.225 in. $\genfrac{}{}{0pt}{}{+.000 \text{ in.}}{-.008 \text{ in.}}$, what percentage of the parts will be acceptable?

a) 85.1%
b) 87.7%
c) 92.7%
d) 98.3%

9. Find the relative maximum and minimum points for the following equation.

$$f(x) = 2x^3 + 3x^2 + 4$$

a) maximum at $x = -1$, minimum at $x = 0$
b) maximum at $x = 1$, minimum at $x = 0$
c) maximum at $x = 2$, minimum at $x = 2$
d) maximum at $x = 2$, minimum at $x = -2$

10. Find the area between the $x$ axis and the function given below between $x = 1$ and $x = 10$.

$$f(x) = 3x^2 + 2x + 1$$

a) 1,007
b) 1,107
c) 1,310
d) 1,550

# Physics and Engineering Sciences

11. How many microns is the approximate equivalent to 0.00020 in.?

12. A 20-in. diameter barrel is 48 in. tall. A person wants to fill it with machining coolant. The mixing ratio of water to coolant concentrate is 3:1. Assuming the coolant concentrate comes in 1-gal increments and no fractions of a gallon can be used, how much coolant concentrate should be purchased to fill the barrel as much as possible?

a) 0.0254 microns
b) 5.08 microns
c) 7.41 microns
d) 12.54 microns

a) 13 gal
b) 16 gal
c) 48 gal
d) 65 gal

13. Which of the following statements is TRUE?

    a) The angle of refraction is the same as the angle of incidence.
    b) Reflected light is the portion of light absorbed by an object.
    c) When a ray of light is transmitted through three materials, its line of travel is unchanged.
    d) Reflected light is the portion of light that bounces off an object.

14. If the sound intensity is $2 \times 10^{-6}$ W/m², find the relative intensity level.

    a) 5.3 dB
    b) 6.3 dB
    c) 53 dB
    d) 63 dB

15. Find the current flowing through the circuit shown in Figure Q15.

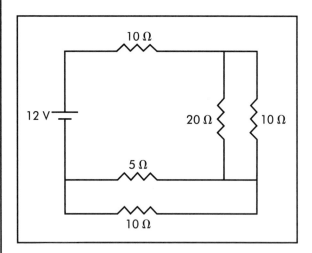

Figure Q15.

    a) 0.3 A
    b) 0.6 A
    c) 1.0 A
    d) 1.2 A

16. If a circuit's voltage source is 20 V and the current is 1.5 A, what is the power sourced from the battery?

17. The power consumed by the resistive load for the circuit in Figure Q17 is _____.

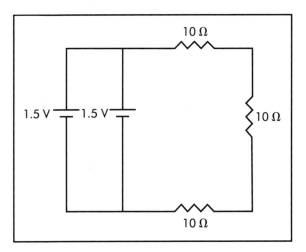

Figure Q17.

a) 1.5 W
b) 13 W
c) 25 W
d) 30 W

a) 0.75 A
b) 0.30 A
c) 75 mA
d) 3 mA

18. For the problem shown in Figure Q18, the tension in the cable is _____ .

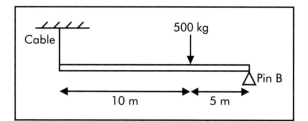

Figure Q18.

19. Referring to Figure Q19, find the magnitude of force, $P$, required to keep the 500-lb block stationary if the coefficient of static friction is 0.25. ($P$ is parallel to the inclined surface.)

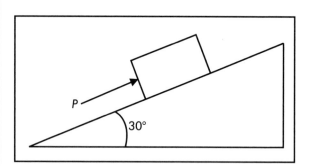

Figure Q19.

a) 167 N
b) 1,635 N
c) 2,453 N
d) 5,027 N

a) 142 lb
b) 166 lb
c) 433 lb
d) 500 lb

20. In Figure Q20, 100 lb is being applied to the end of a wrench. How much torque is being applied to the bolt?

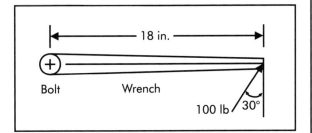

Figure Q20.

21. A worker drops a wrench down the elevator shaft of a tall building. How far will the wrench fall in 1.5 seconds?

a) 900 in.-lb
b) 1,559 in.-lb
c) 1,800 in.-lb
d) 2,459 in.-lb

a) 3 m
b) 5 m
c) 7 m
d) 11 m

## Physics and Engineering Sciences

22. A flywheel is rotating at 200 rpm. It is brought to a stop at a rate of 2 rad/s². Approximately how many revolutions will the wheel make from the time it begins decelerating until it comes to a complete stop?

23. Find the force required to stretch a 12-in. long, 0.500-in. diameter, low-carbon steel rod 0.010 in. $E_{steel} = 30 \times 10^6$ psi.

a) 13 revolutions
b) 18 revolutions
c) 21 revolutions
d) 30 revolutions

a) 3,680 lb
b) 4,540 lb
c) 4,909 lb
d) 5,150 lb

24. A 2-m long, low-carbon steel rod is subjected to a tensile load of 12 kN. The allowable stress is not to exceed 50 MPa. Calculate the required rod diameter.

25. A 2-ft long, hollow aluminum shaft has an outside diameter of 2 in. and an inside diameter of 1.5 in. Calculate the angle of twist in the 2-ft length when a torque of 5,000 in.-lb is applied.

$G_{aluminum} = 3.9 \times 10^6$ psi

a) 10 mm
b) 18 mm
c) 33 mm
d) 65 mm

a) 0.002 rad
b) 0.012 rad
c) 0.020 rad
d) 0.029 rad

26. A quantity of gas is contained in an insulated cylinder with a perfectly sealing piston. If the gas does 10 J of work on the piston and the change in internal energy is 0, how much energy must have been added to the gas?

   a) 0 J
   b) 5 J
   c) 10 J
   d) 20 J

27. In heat treating, a hot steel part being quenched with coolant is an example of which type of heat transfer?

   a) convection
   b) radiation
   c) conduction
   d) conduction and radiation

28. A 12-in. diameter steel shaft must be removed from a machine. The shaft needs to shrink 0.005 in. in diameter to be removed. Assuming the shaft starts at room temperature (72° F [22° C]), what final temperature will it need to be to shrink 0.005 in. in diameter? (For the steel shaft, use the coefficient of expansion for iron.)

29. Water is flowing through a 1-in. diameter supply line at 5 gal/min. The 1-in. supply line is reduced to a 0.5 in. diameter discharge line. What is the discharge velocity of the water in in./min?

a) 13° C
b) 7° F
c) 0° F
d) –13° C

a) 924 in./min
b) 2,924 in./min
c) 3,650 in./min
d) 5,882 in./min

30. Pressure of 100 psi is exerted on the 3-in. diameter side of the sealed piston in Figure Q30. How much force can the 1-in. diameter piston rod create?

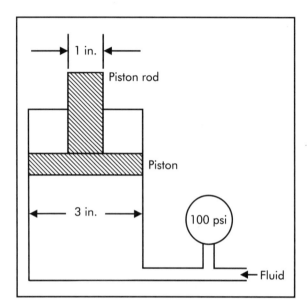

Figure Q30.

a) 225 lb
b) 510 lb
c) 707 lb
d) 1,107 lb

# Materials

31. What is the minimum quench temperature at which plain carbon steel forms martensite? Assume the initial temperature is 800° C.

    a) 110° C
    b) 220° C
    c) 30° C
    d) 70° C

32. The minimum temperature for full annealing 1040 steel is closest to _____.

    a) 400° C
    b) 600° C
    c) 800° C
    d) 900° C

33. Which of the following is a TRUE statement regarding cold working?

    a) The effects of cold working are more noticeable in ferrous metals.
    b) Cold working is performed above the recrystallization temperature.
    c) Cold working increases a metal's ductility.
    d) Cold working increases a metal's hardness.

34. A higher percentage of elongation for a given material indicates _____.

    a) lower ductility
    b) higher hardness
    c) higher ductility
    d) higher tensile strength

Materials

35. Which of the following alloying elements is typically added to steel used for steam lines to improve creep resistance?

   a) chrome
   b) nickel
   c) vanadium
   d) molybdenum

36. Which of the following plastics is the most dense?

   a) phenolic
   b) polystyrene
   c) polypropylene
   d) polyethylene

37. Crystallinity in plastics creates _____.

   a) higher transparency and lower strength
   b) lower strength and lower density
   c) lower density and lower transparency
   d) higher strength and higher density

38. The following groups of materials are thermoplastics EXCEPT _____.

   a) polyvinyl-chloride and polystyrene
   b) polyethylene, polystyrene, and polypropylene
   c) polystyrene, phenolic, and nylon
   d) nylon, acetal, and high-density polyethylene

39. In a composite material _____.

   a) the orientation of the fibers does not affect the overall strength of the composite
   b) the fibers and matrix material dissolve together to form a solid solution
   c) much of the overall strength comes from the fibers rather than the matrix
   d) it is important that the matrix material remains stiff and does not deform

40. A ceramic may be chosen for a specific application if _____.

   a) high fatigue strength and low thermal conductivity are required
   b) low density and high temperature resistance are required
   c) low density and high fatigue strength are required
   d) high temperature resistance and corrosion resistance are required

# Product Design

41. A basic size followed by a plus and minus expression where variations are allowed in only one direction from the nominal is called _____.

    a) limit dimensioning
    b) bilateral tolerance
    c) allowance dimensioning
    d) unilateral tolerance

42. Which of the following is the maximum material condition (MMC) of a hole with a dimension of ⌀0.625 in. ±0.005 in.?

    a) 0.625 in.
    b) 0.630 in.
    c) 0.620 in.
    d) 0.615 in.

43. Tighter (smaller) tolerances _____.

    a) force manufacturing to make perfect parts
    b) lead to lower manufacturing costs
    c) reduce the number of engineering change orders
    d) lead to higher manufacturing costs

44. A shaft is dimensioned 0.990 in. ± 0.002 in. and the matching bore is dimensioned 1.000 in. ± 0.004 in. The allowance is _____.

    a) 0.004 in.
    b) 0.006 in.
    c) 0.010 in.
    d) 0.016 in.

45. Find the largest assembly clearance for a 1-in. diameter hole and shaft with an RC 6 fit.

    a) 0.0016 in.
    b) 0.0048 in.
    c) 0.0078 in.
    d) 0.0155 in.

46. The part with the smoothest surface finish would be the part with an average roughness of _____.

    a) 10 µin.
    b) 50 µin.
    c) 100 µin.
    d) 150 µin.

47. Referring to Figure Q47, the maximum value of $X$ is _____.

   a) 3.0 in.
   b) 3.2 in.
   c) 3.3 in.
   d) 3.5 in.

48. When only a tolerance of size is specified, the part's form and size are controlled by _____.

   a) Rule 1
   b) Rule 2
   c) title block tolerance
   d) ANSI standards

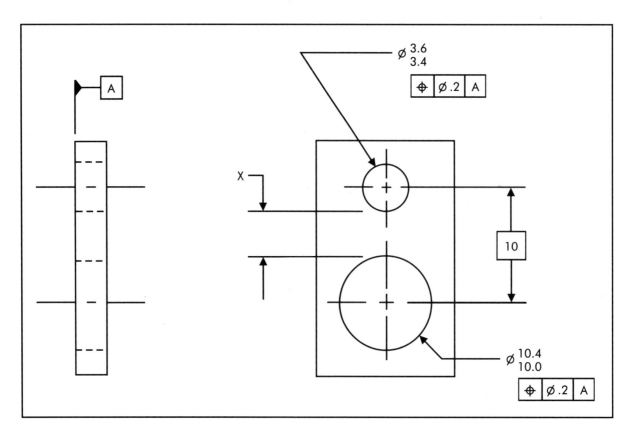

Figure Q47.

49. Which of the following specifies the type, size, and shape of the geometric tolerance zone, dictates datums, and assigns material condition modifiers?

a) detailed dimensions and notes
b) datum reference frame
c) operations sheet
d) feature control frame

50. The symbol ⌀ before a tolerance value indicates _____.

a) the tolerance zone it applies to is cylindrical
b) the tolerance only applies if the part is circular
c) the tolerance zone it applies to is concentric
d) the tolerance increases as the part departs from the maximum material condition (MMC)

51. Referring to Figure Q51, what is the maximum permissible flatness error of surface *A*?

a) 0.0 mm
b) 0.1 mm
c) 0.2 mm
d) 0.4 mm

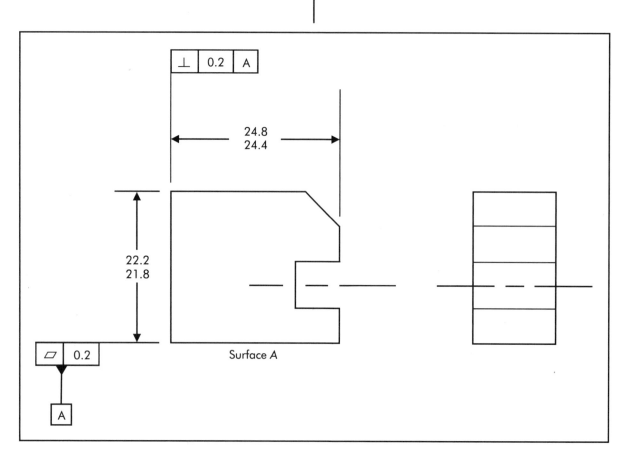

Figure Q51.

52. Referring to Figure Q52, the gage for verifying the straightness control would be a pin that is _____ in diameter.

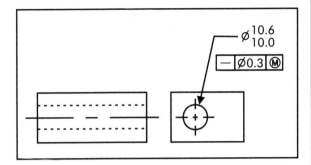

Figure Q52.

a) 9.7 mm
b) 10.0 mm
c) 10.6 mm
d) 10.9 mm

53. Referring to Figure Q51, what is the shape of the perpendicularity control tolerance zone?

a) two parallel lines
b) two parallel planes
c) a cylinder
d) two perpendicular planes

54. In GD&T, which type of geometric control can control the relationship of one or more features to a datum axis?

a) flatness
b) straightness
c) runout
d) circularity

55. What type of CAD model can be used to calculate quantities such as weight, moment of inertia, and center of gravity?

a) wireframe model
b) surface model
c) IGES model
d) solid model

56. Which of the following is TRUE about rapid prototyping?

a) Rapid prototyping helps detect early design flaws.
b) Rapid prototyping can produce surface finishes similar to those produced by precision grinding.
c) Parts produced by rapid prototyping are typically used for physical testing of properties such as fatigue and tensile strength.
d) Rapid prototyping contributes to the part realization process and the expense can not usually be justified.

Product Design

57. The process of "picking points" off of a part surface to generate a 3D-CAD model is known as ____.

 a) selective laser sintering
 b) wireframe modeling
 c) stereolithography
 d) reverse engineering

58. Which of the following CAD file formats is typically used for rapid prototyping?

 a) STL
 b) CSG
 c) IGES
 d) BREP

59. According to design for assembly (DFA) methodology, two components can be produced as a single component if ____.

 a) they do not move relative to each other
 b) they need to be different materials
 c) they must move relative to each other
 d) for disassembly purposes the parts need to be separate

60. When performing a design failure mode and effects analysis (FMEA) for a tire, vibration from abnormal wear would be considered a ____.

 a) failure mode
 b) failure effect
 c) causes of a failure
 d) random error

61. Which of the following is TRUE about quality function deployment (QFD)?

 a) QFD is a single-step process.
 b) QFD is driven by the voice of the customer.
 c) QFD decreases product design time.
 d) For QFD to work properly, customers should be technically literate.

62. Group technology ____.

 a) increases design time due to part classification requirements
 b) takes advantage of the uniqueness of each part design
 c) can reduce part proliferation and redundant designs
 d) increases manufacturing costs due to part coding

63. With respect to group technology, attribute coding _____.

   a) uses a sequence of digits whereby the subsequent digit is dependent on the preceding digit
   b) requires the same number of digits in the code as there are unique part attributes
   c) is based on operation and routing sheets
   d) can not be used with castings, forgings, and injection-molded parts

64. With regard to a failure mode and effects analysis (FMEA), _____.

   a) the RPN can never exceed 100
   b) lower RPN values signal problems that require immediate attention
   c) the RPN can never be lower than 5
   d) high RPN values signal problems

65. Measurement of how good a product design is with respect to an assembly is calculated by using _____.

   a) design for assembly
   b) economic order quantity
   c) risk priority number
   d) design efficiency

# Manufacturing Processes

66. Using the Taylor equation for tool wear, Let $n = 0.125$, $V=100$ ft/min, and $C = 200$ ft/min ($C$ is the cutting speed at $T = 1$). Find the tool life.

   a) 256 min
   b) 156 min
   c) 426 min
   d) 96 min

67. Choosing from the list below, which cutting tool material has the highest wear resistance combined with poor impact strength?

   a) high-speed steel
   b) coated carbide
   c) ceramic
   d) diamond

## Manufacturing Processes

68. Calculate the maximum feed rate in a turning operation using the following parameters: 4-in. round, 1020 steel, 0.060-in. depth of cut, 100-ft/min cutting speed, 2-horsepower motor, 80% spindle efficiency, and a unit horsepower of 1.

    a) 0.212 in./rev
    b) 0.360 in./rev
    c) 0.048 in./rev
    d) 0.022 in./rev

69. The most accurate lathe chuck for turning round stock is the _____.

    a) lathe dog
    b) collet
    c) three jaw
    d) face plate

70. Calculate the drill rpm for a 1-in. diameter drill cutting material at a speed of 100 ft/min.

    a) 182 rpm
    b) 282 rpm
    c) 382 rpm
    d) 482 rpm

71. Calculate the maximum depth of cut for a four-flute, 1-in. end mill using the following parameters: a material cutting speed of 100 ft/min, a feed rate of 4 in./min, a 3-horsepower motor, and 75% spindle efficiency. The width of cut is 1 in. and the unit horsepower is 1.

a) 0.566 in.
b) 0.256 in.
c) 0.466 in.
d) 0.356 in.

72. When compared to up-milling, climb (down) milling generally _____.

a) requires more feed force
b) increases the cutter temperature
c) decreases the surface finish quality
d) requires less feed force

73. In bandsawing, the number of teeth per inch is called the _____.

a) pitch
b) set
c) kerf
d) hook

74. All of the following are true about cold forming, EXCEPT _____.

a) cold forming is performed below the recrystallization temperature
b) cold-formed parts are strain hardened
c) cold-formed parts have a poor surface finish
d) cold-formed parts have good dimensional accuracy

75. Why is a forged part stronger than a cast part?

a) Forged parts typically have a better surface finish.
b) Forged parts are generally made from low-melting-temperature alloys.
c) Forged parts are susceptible to porosity defects.
d) Forged parts have their grain aligned to the contour of the part.

## Manufacturing Processes

76. Which of the following processes can produce great detail with only a small amount of deformation?

    a) heading
    b) extrusion
    c) hemming
    d) coining

77. For a 2 × 2 in. square steel blank, calculate the maximum material thickness that can be blanked with a 40-ton press if the material's shear strength is 20,000 psi.

    a) 0.200 in.
    b) 0.300 in.
    c) 0.400 in.
    d) 0.500 in.

78. Completely folding over the edge of a piece of sheet metal to provide strength and rigidity, such as on the edges of an automobile door, is known as _____.

    a) flanging
    b) chamfering
    c) extruding
    d) hemming

79. Generating holes in sheet metal without completely removing the metal is known as _____.

    a) piercing
    b) lancing
    c) notching
    d) parting

80. In powdered metals, binders are added to the powder to _____.

    a) give the part more strength after sintering
    b) increase part density
    c) reduce infiltration during sintering
    d) increase part strength after compaction

81. In powder metallurgy, which stage of the sintering cycle fuses the particles together?

    a) 1st
    b) 2nd
    c) 3rd
    d) 4th

## Manufacturing Processes

82. In casting, what is the part of the gating system that leads directly into the mold cavity called?

    a) sprue
    b) gates
    c) runners
    d) pouring cup

83. Which process provides the highest quality casting (that is, the best surface finish, dimensional accuracy, integrity, etc.)?

    a) sand casting
    b) lost foam casting
    c) vacuum casting
    d) investment casting

84. Which of the following welding processes forms a protective slag layer over the weld?

    a) SMAW
    b) resistance spot welding
    c) GTAW
    d) gas fusion welding

85. Which of the following is an advantage of electrical resistance spot welding over stick welding?

    a) It can weld thick metals.
    b) It can weld dissimilar metals.
    c) No filler metal or flux is required.
    d) Electrodes are consumable.

86. The welding process that uses high-frequency vibration and pressure to join metals is known as _____.

    a) MIG welding
    b) resistance welding
    c) friction welding
    d) ultrasonic welding

87. The electric charge applied to the paint particles in electrostatic spraying _____.

    a) decreases the overall transfer efficiency due to the electric supply required
    b) decreases the possible coating thickness
    c) increases the amount of paint required due to the inverse magnetic field
    d) increases the transfer efficiency due to the magnetic attraction

88. The process of electroplating requires _____.

    a) an insulated workpiece
    b) a direct current (DC) power source
    c) an inert atmosphere
    d) an air pressure source

89. An automotive plastic gas tank is most likely manufactured by _____.

   a) blow molding
   b) injection molding
   c) thermoforming
   d) casting

90. Which of the following processes is used to produce plastic tubing?

   a) rotational molding
   b) injection molding
   c) blow molding
   d) extrusion

91. The process for producing long lengths of molding with a consistent cross-section made from a composite material would most likely be _____.

   a) resin transfer molding (RTM)
   b) rolling
   c) pultrusion
   d) filament winding

92. Ceramic parts used in engineering applications are generally made by _____.

   a) extrusion
   b) isostatic pressing
   c) jiggering
   d) casting

93. In slip casting, _____.

   a) the part's surface finish is determined by the amount of time the slip is in the plaster mold
   b) the amount of part glazing is determined by the amount of time the slip is in the plaster mold
   c) the part's wall thickness is determined by the amount of time the slip is in the plaster mold
   d) the firing temperature is determined by the amount of time the slip is in the plaster mold

94. The process for making traces on a printed circuit board is known as _____.

   a) screen printing
   b) wave soldering
   c) stencil printing
   d) selective etching

95. Surface-mounted devices (SMD) _____.

   a) can not be used on the same board with through-hole technology devices
   b) are soldered in place using reflex soldering
   c) are soldered in place using wave soldering
   d) are temporarily secured by solder paste before soldering

# Production Systems

96. A customer places an order at a local pizzeria where pizzas are not made in advance. The pizzeria is an example of which of the following production environments?

    a) manufacture-to-stock
    b) assemble-to-order
    c) design-to-order
    d) engineer-to-order

97. Forecasting becomes more accurate when ____.

    a) a single forecasting model is used to eliminate conflicting information from other models
    b) the forecast period gets longer showing the bigger picture
    c) the forecast period is short
    d) set-based programming is used rather than sequential programming

98. Which of the following is NOT true about the master production schedule?

    a) It is an input to a material requirements planning (MRP) system.
    b) It determines where bottlenecks will occur in the manufacturing process.
    c) It is updated every 6-12 months.
    d) It translates the aggregate plan into a separate plan for individual items.

99. Which of the following is NOT an element of MRP?

    a) bills of materials
    b) process routings
    c) customer orders
    d) personnel data

## Production Systems

100. How does capacity requirements planning differ from rough-cut capacity planning?

    a) It determines where bottlenecks will occur.
    b) It considers work-in-process and replacement parts.
    c) It is less detailed than rough-cut capacity planning.
    d) It does not consider work-in-process or replacement parts.

101. The scheduling technique that is a function of lead time and due date is known as _____.

    a) first-in-first-out (FIFO)
    b) start date priority
    c) due date priority
    d) critical ratio

102. Material requirements planning (MRP) is a _____.

    a) rate-based material planning system
    b) system of kanbans and reorder points
    c) time-phased material planning system
    d) pull-based material planning system

103. How does manufacturing resource planning (MRPII) differ from MRP?

    a) MRPII is a pull system and controls finished goods.
    b) MRPII accounts for work-in-process and original equipment manufacturer (OEM) replacement parts.
    c) MRPII plans and schedules all the resources in a manufacturing company.
    d) MRPII does not include work-in-process and OEM replacement parts.

104. Which of the following is TRUE about a pull system?

    a) It only produces parts as required by the subsequent operation.
    b) It replaces just-in-time (JIT) manufacturing.
    c) It is a component of MRP.
    d) It only works in a low-volume production environment.

## Production Systems

105. Lean production differs from mass production in that lean production _____.

    a) produces cost-competitive products due to fewer varieties of products and high volume
    b) utilizes unskilled workers with a wide variety of high-speed fixed automation
    c) utilizes multi-skilled workers and flexible equipment to produce a variety of cost-competitive products
    d) sets an acceptable number of defects based on infrequent changeovers

106. A visual cue that signals the need for stock replenishment is known as a(n) _____.

    a) MRP
    b) kaizen
    c) standardization
    d) kanban

107. In lean production, safety stock _____.

    a) is acceptable
    b) is not counted on the value stream map
    c) is counted as work-in-process (WIP)
    d) is continually reduced

108. Call lights, also known as andon boards, are examples of _____.

    a) production scheduling
    b) visual control
    c) material requirements planning (MRP) inputs
    d) 5S (sort, straighten, shine, standardize, sustain)

109. The process of reducing overtime and ensuring the proper utilization of workers while at the same time ensuring constant flow to and from suppliers is known as _____.

    a) production leveling
    b) line balancing
    c) production flow analysis
    d) process planning

110. Installing a vision system to detect missing parts on an assembly prior to shipping is an example of _____.

    a) error proofing
    b) mistake proofing
    c) 5S
    d) standardization

Production Systems

111. During an 8-hour workday, workers have 30 min for lunch and two 15-min breaks. Immediately at the end of the workday a truck ships the products to the customer. The customer requires 600 cases/day at 10 parts/case. What is the takt time?

a) 75 cases/hour
b) 0.08 min/part
c) 0.07 min/part
d) 0.07 cases/hour

112. The process of graphically demonstrating the value-added and non-value-added activities from raw materials to the finished product and then to the customer is known as _____.

a) master production scheduling
b) forecasting
c) operation sheet analysis
d) value stream analysis

113. The process or program whereby the plant floor is organized, employees put tools and materials where they belong, and equipment is kept clean is known as _____.

a) 8D
b) industrial hygiene
c) 5S
d) best practices

114. Just-in-time manufacturing focuses on all of the following EXCEPT _____.

a) reducing the number of rejected parts
b) reducing setup times
c) reducing excessive inventory levels
d) reducing workers' responsibility for quality

115. During the part design phase and after the part design is finalized, the process of determining which manufacturing processes are to be used, the tooling requirements, and developing routing and operations sheets is known as _____.

a) capacity planning
b) process planning
c) production planning
d) material requirements planning

116. Which type of facility layout groups together similar functions?

a) process layout
b) product-process layout
c) fixed layout
d) synchronous layout

117. The average time to assemble one lock is 60 sec. The percent rating of the operator is 75% and the allowance for material delays is 5%. Calculate the standard time.

a) 42.4 sec
b) 45.0 sec
c) 47.3 sec
d) 50.4 sec

118. Which assembly process is used when the time required for performing different operations varies greatly?

a) synchronous
b) nonsynchronous
c) continuous
d) rotary

119. How do jigs differ from fixtures?

a) Jigs are workholders and fixtures are not.
b) Jigs guide the cutting tool and fixtures do not.
c) Jigs use clamps while fixtures use locators.
d) Jigs do not position the workpiece as accurately as fixtures do.

120. Which type of fixture would be used to machine equally spaced holes for a bolt circle?

a) 3-2-1 fixture
b) vise-jaw fixture
c) collet fixture
d) indexing fixture

121. An expert computer system that can determine the machinery needed, sequence of operations, and tooling to produce a new part is known as _____.

a) FMEA
b) CAD
c) MRP
d) CAPP

122. How many points are required to completely restrict the movement of a part with respect to three degrees of freedom?

a) 3
b) 6
c) 9
d) 12

123. Using load cells to measure forces at various points on a press during each stamping cycle and then comparing the measurements in real time to a nominal or acceptable value is an example of _____.

a) predictive maintenance
b) capacity planning and measurement
c) methods engineering
d) takt time measurement

124. A plant with excess physical space allows its suppliers to manufacture their respective parts in the same physical building as seen in Figure Q124. Which type of facility layout is being used?

| Rotors | Spindles |
|---|---|
| Bearings | Axle assembly |

Figure Q124.

a) process layout
b) product-process layout
c) fixed layout
d) semi-fixed layout

125. A diamond pin locator is an example of a(n) _____.

   a) relieved locator
   b) method of locating a planar surface
   c) integral locator
   d) alternative to the 3-2-1 principle

126. In designing a fixture, clamps should be located _____.

   a) over a locator if possible
   b) on the bottom of the part
   c) as close together as possible
   d) on the thinnest sections of the part

127. The public's demand for bottled water has nearly doubled. The demand for plastic bottles to package the water has also increased. The plastic bottle demand is known as a(n) _____.

   a) independent demand
   b) related demand
   c) dependent demand
   d) coincidental demand

128. According to ABC inventory management, _____.

   a) "A" items are reviewed less frequently than "C" items
   b) in terms of dollars, 80% of the items will be involved in 20% of the usage
   c) "C" items are reviewed more frequently than "B" items
   d) the quantity on hand is the same for all items in inventory

129. A company has monthly inventory usage of $10,000. Its average inventory is $5,000. Find the number of annual inventory turns.

   a) 2
   b) 12
   c) 24
   d) 30

130. A low number of inventory turns _____.

   a) is better than a high number of inventory turns
   b) indicates a supplier carries a small amount of inventory and replenishes it frequently
   c) indicates a supplier carries a large amount of inventory and replenishes it infrequently
   d) is a result of lean manufacturing

## Production Systems

131. Calculate the economic order quantity (EOQ) if the annual usage is 100,000 units, setup and order costs are $25, the unit cost is $10, and there is an interest and storage cost of 8%.

    a) 250
    b) 791
    c) 2,500
    d) 625,000

132. Which acronym identifies a professional society that specializes in inventory management?

    a) SME
    b) ASME
    c) ASQ
    d) APICS

133. Coordinating the flow of materials, information, and money between customers and suppliers is known as _____.

    a) supply chain management
    b) inventory management
    c) value stream management
    d) value-added management

134. According to ABC inventory management, machine screws would be considered _____.

    a) "A" items
    b) "B" items
    c) "C" items
    d) either "A" or "C" items

135. Near-zero inventory on purchased materials can be achieved partially by _____.

    a) ordering larger lots, thereby eliminating the effect of long lead times
    b) focusing on short-term costs and expediting
    c) working with suppliers to produce zero defects
    d) increasing the overall number of suppliers in case of an emergency

# Automated Systems and Control

136. Justification for the implementation of automation should include_____.

    a) how automation will improve the quality of work life and level of technology in the plant
    b) how automation will improve manufacturing productivity
    c) how automation will affect the preventive maintenance program and increase overtime
    d) how automation will improve the quality of work life and increase the amount of labor required

137. The software that connects the user with the network is called_____.

    a) network interface card
    b) network bridge
    c) network Ethernet adapter card
    d) network operating system

138. What are the rules for communicating data in a network called?

    a) network interface
    b) architecture
    c) protocol
    d) network topology

139. Milling the letter "B" would require _____.

    a) circular interpolation
    b) linear interpolation
    c) point-to-point control
    d) linear and circular interpolation

## Automated Systems and Control

140. Which letter will the program in Figure Q140 generate?

```
N10 G00 X0 Y0. T4M6
N20 G00 X1 Y1
N30 G01 Z-.1 F160 S4000
N40 G01 X1 Y3
N50 G02 X1 Y2 I1 J2.5
N60 G01 X2 Y1
N70 G00 Z2
```

Figure Q140.

a) X
b) C
c) A
d) R

141. A household heating system that includes a furnace and a thermostat is an example of a(n) _____.

a) open-loop system
b) PID system
c) resolver system
d) closed-loop system

142. Consider the ladder logic program in Figure Q142. Contacts A1-1 and B1-1 are normally open. Contact D1-1 is normally closed. Output C1 represents a light bulb. If the push-button switch is depressed and *held* in the depressed position, output A1 is energized, thereby closing the contact A1-1. If B1-1 remains off (open contact), what is the state of output C1?

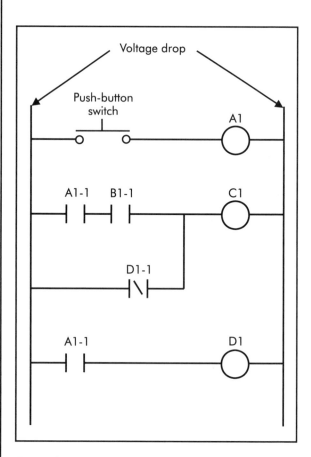

Figure Q142.

a) on
b) off
c) blinking continuously
d) negative

Automated Systems and Control

143. Continuous path control of a robot requires _____.

 a) adaptive control
 b) multiple axes controlled simultaneously
 c) point-to-point control
 d) a teach pendant for programming

144. A robotic application for dispensing silicone gasket material on mating parts prior to assembly would most likely use the _____.

 a) part coordinate system
 b) world coordinate system
 c) universal coordinate system
 d) polar coordinate system

145. A common camera used in machine vision is the _____.

 a) PTA
 b) FDM
 c) SLR
 d) CCD

# Quality

146. All of the following are part of the total quality management (TQM) philosophy EXCEPT _____.

    a) detecting defects is more important than preventing them
    b) continuous improvement never ends
    c) improvements should be linked to business metrics
    d) the organization must be customer-driven

147. The recall of Firestone's Wilderness® AT tires is an example of a(n) _____.

    a) internal quality cost
    b) external quality cost
    c) prevention quality cost
    d) appraisal quality cost

148. ISO 9000 registration _____.

    a) verifies less than 3.4 parts-per-million defective from a supplier
    b) verifies a supplier is lean
    c) verifies a supplier has high process capability
    d) verifies a supplier has a quality system in place

149. A company-wide focus on providing quality service is known as _____.

    a) total value analysis and engineering
    b) quality function deployment
    c) total quality management
    d) total quality control

150. Which of the following statements is TRUE regarding the Malcolm Baldrige Award?

    a) Leadership is the highest weighted Baldrige Award criteria.
    b) ISO 9000 registration does not automatically qualify a company for the Baldrige Award.
    c) The Baldrige Award is similar to ISO 9000 in that it also focuses on outcomes and customer satisfaction.
    d) The Baldrige Award can only be given to manufacturers.

# Quality

151. The maximum torque generated by a 12-volt cordless drill is an example of _____.

    a) variable data
    b) continuous data
    c) assignable cause data
    d) attribute data

152. A supplier casts six-cylinder diesel engine blocks and wants to monitor/chart the number of voids per casting. The appropriate control chart to use would be a(n) _____.

    a) $x$-bar and $R$ chart
    b) $c$ chart
    b) $u$ chart
    d) $p$ chart

153. A chipped cutting tool would create _____.

    a) unusable data
    b) high process capability
    c) assignable variability
    d) a low standard deviation

154. Inspection is _____.

    a) a value-added activity
    b) a non-value-added activity
    c) required for good process capability
    d) a cost-cutting activity

155. Calculate the upper control limit for an $x$-bar chart given the data in Table Q155.

Table Q155.

| Subgroup | Shaft Diameters (mm) | | | |
|---|---|---|---|---|
| | A | B | C | D |
| 1 | 25 | 30 | 23 | 28 |
| 2 | 23 | 24 | 25 | 22 |
| 3 | 31 | 29 | 25 | 25 |
| 4 | 25 | 26 | 27 | 25 |
| 5 | 30 | 29 | 25 | 27 |

   a) 22.9
   b) 27.1
   c) 28.9
   d) 29.6

156. Calculate the lower control limit for an $R$ chart using the data from Question 155.

    a) 0
    b) 4.6
    c) 9.1
    d) 9.7

157. Assume the tolerance on the dimension being measured is 20 mm ±2 mm. Find the process capability using the data in Table Q157.

    Table Q157.

    | Subgroup | Part Thickness (mm) | | |
    |---|---|---|---|
    | | A | B | C |
    | 1 | 20 | 19 | 20 |
    | 2 | 21 | 20 | 20 |
    | 3 | 20 | 19 | 19 |
    | 4 | 21 | 20 | 20 |

    a) 0.6
    b) 1.1
    c) 1.2
    d) 1.4

158. A production run of 1,000 shafts with a nominal diameter of 0.504 in. has a mean of 0.500 in. and standard deviation of 0.002 in. If the shaft diameters are normally distributed, how many shafts will be smaller than 0.504 in.?

   a) 478
   b) 500
   c) 956
   d) 978

159. To increase process capability, _____.

   a) reduce variation
   b) reduce the part tolerance
   c) increase variation
   d) reduce the part tolerance and increase variation

160. A high $C_p$ and low $C_{pk}$ means _____.

   a) the process is out of control
   b) the process has high variability and the process distribution is centered with respect to the nominal
   c) the process has low variability and the process distribution is centered with respect to the nominal
   d) the process has low variability and the process distribution is skewed either to the left or right of the nominal

161. Acceptance sampling plans _____.

   a) guarantee all the lots or shipments accepted from a supplier are good
   b) require the consumer to risk rejecting bad parts
   c) are not required if suppliers ship zero defects
   d) ensure that all parts in a lot are acceptable based on a sample

162. A passenger riding in the front seat of a car can not accurately read the fuel gage due to which type of error?

   a) parallax error
   b) bias error
   c) technique error
   d) operator error

## Quality

163. Design a go/no-go gage capable of qualifying a shaft with the following diameter: 1.000 in. ± 0.010 in. Use a 10% gage tolerance and 5% wear allowance.

164. Which of the following is a destructive test?

    a) tensile test
    b) magnetic particle inspection
    c) ultrasonic inspection
    d) liquid penetrant inspection

165. Over an 8-hour shift an operator measures 100 shaft diameters and gets the exact same diameter for all 100 shafts. The probable reason for this is _____.

    a) the measurement instrument lacks precision
    b) the measurement instrument is not accurate
    c) the measurement instrument is not repeatable
    d) the manufacturing process is extremely capable and all 100 shafts have the exact same diameter

166. Qualification of internal diameters can be done with a _____.

    a) snap gage
    b) optical comparator
    c) rule
    d) plug gage

a) go side = $1.010 \, ^{+.000}_{-.002}$, no-go side = $.990 \, ^{+.002}_{-.000}$

b) go side = $1.009 \, ^{+.002}_{-.000}$, no-go side = $.990 \, ^{+.000}_{-.002}$

c) go side = $1.009 \, ^{+.000}_{-.002}$, no-go side = $.989 \, ^{+.002}_{-.000}$

d) go side = $1.009 \, ^{+.000}_{-.002}$, no-go side = $.990 \, ^{+.002}_{-.000}$

167. What is the diameter of the shaft measured by the 0–1.0-in. micrometer in Figure Q167?

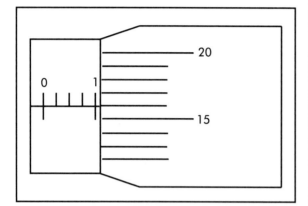

Figure Q167.

a) 0.114 in.
b) 0.116 in.
c) 0.151 in.
d) 1.160 in.

168. Which of the following is TRUE about dial indicators?

a) As the dial indicator's range increases, the accuracy also increases.
b) Highly sensitive dial indicators generally have a high degree of amplification.
c) If the part tolerance is ±0.010 in., the dial indicator discrimination should be approximately 0.004 in.
d) Dial indicators measure based on comparison as opposed to measuring directly.

169. The device for measuring small and/or complex parts, which projects a magnified image on a screen, is known as a _____.

a) coordinate measuring machine (CMM)
b) profilometer
c) vernier caliper with digital display
d) optical comparator

170. The process of combining gage blocks is known as _____.

a) diffusing
b) wringing
c) adhering
d) joining

# Manufacturing Management

171. In the line-and-staff organization, _____.

   a) information flows horizontally across departments rather than vertically
   b) channels of communication are well established
   c) staff relationships are considered supervisory in nature
   d) line relationships are advisory in nature

172. Allowing individual departments or groups to do their own purchasing for items that cost less than $500 is known as _____.

   a) goal setting
   b) ABC management
   c) decentralization
   d) span of control

173. Pareto analysis can _____.

   a) identify cause-and-effect relationships
   b) distinguish the trivial few from the vital many
   c) identify the root cause of a problem
   d) determine key process variables to chart or monitor

174. Good teamwork is the result of many factors including _____.

   a) close supervision
   b) positive reinforcement
   c) subjective team-member evaluation
   d) abstract and enticing goals

175. According to participatory management, _____.

   a) employees are empowered to implement their decisions
   b) workers are generally lazy and need constant supervision
   c) labor is a commodity
   d) management solicits input from individuals if necessary

## Manufacturing Management

176. A company can acquire protection for one of its brand names through a _____.

    a) patent
    b) trade secret
    c) trademark
    d) copyright

177. Collective bargaining _____.

    a) obligates one side to concede or agree to the other side's proposal
    b) requires both parties to meet in good faith and resolve their differences within 30 days prior to the expiration of the current contract
    c) permits labor to strike prior to the end of the current contract
    d) permits labor to strike after 60 days of notice to management

178. The type of company where employees are required to join the union and bargain with the company collectively is a(n) _____.

    a) open shop
    b) union shop
    c) agency shop
    d) exclusive bargaining shop

179. The type of fire extinguisher that can be used on electrical equipment fires is _____.

    a) "A"
    b) "B"
    c) "C"
    d) "D"

180. Manufacturing safety data sheets (MSDS) _____.

    a) are not required if a company has 10 or fewer employees
    b) are required only for nonhazardous materials
    c) must be readily available to all employees
    d) can be stored exclusively at the corporate headquarters if the headquarters and all production facilities are located in the same state

181. Tenosynovitis is _____.

    a) an emotional disorder
    b) a repetitive-motion injury
    c) a blood-borne pathogen
    d) a strain of hepatitis B

182. For an assembly station, how high should standing workers be expected to reach according to anthropometric data?

   a) 78.9 in.
   b) 73.0 in.
   c) 90.8 in.
   d) 84.7 in.

183. What is the minimum recommended illumination for an office?

   a) 20 fc
   b) 40 fc
   c) 60 fc
   d) 100 fc

184. Hearing protection is NOT necessary if _____.

   a) continuous exposure at 100 dBA is less than 4 hours
   b) the company does not require hearing protection
   c) noise abatement reduces the noise to an acceptable level
   d) the noise frequency is greater than 1,000 Hz

185. On-the-job safety and health is _____.

   a) the responsibility of employees to ensure
   b) a function of company size
   c) unenforceable by law
   d) achieved if integrated throughout the business or organization

186. To raise money for a new business, a man asks you to loan him some money. He offers to pay you $3,000 at the end of four years. How much should you give him now if you want 12% interest per year on your money?

   a) $1,907
   b) $1,840
   c) $2,106
   d) $1,580

## Manufacturing Management

187. The initial cost of a workstation is $5,000 and its maintenance cost is $500 per year. At the end of its useful 5-year life, its salvage value is $600. Based on an 8% interest rate, calculate the equivalent uniform annualized cost.

a) $1,250
b) $1,450
c) $1,650
d) $1,750

188. Decreasing the number of parts in an assembly will generally _____.

a) increase inventory cost
b) increase direct labor
c) decrease fixed cost
d) decrease variable cost

189. A gear manufacturer can produce 2,000 gears per month at full capacity. The variable cost is $10 per gear, overhead cost is $25,000 per month, and labor cost is $5 per gear. If the gears sell for $30 each, what is the minimum percent capacity the plant needs to use to break even?

a) 50.4%
b) 73.4%
c) 83.4%
d) 96.4%

190. In value engineering and analysis _____.

a) value is defined as the ratio of cost to function (performance)
b) the goal is to eliminate costly required functions
c) teams work to achieve required functions without sacrificing manufacturability
d) esteem value is composed of the product properties that achieve the desired functions

# Personal/Professional Effectiveness

191. A good listener must do all of the following EXCEPT _____.

    a) provide feedback
    b) be motivated
    c) focus on the delivery style
    d) reserve judgment until the end of the message

192. The best format for displaying the number of monthly warranty claims for the past 6 months in a report to the company president would be a _____.

    a) line chart and/or bar chart
    b) table
    c) radar chart
    d) spreadsheet

193. Meeting preparation begins by _____.

    a) developing an agenda
    b) selecting a meeting time and location
    c) determining who will be invited to the meeting
    d) determining the meeting's objectives

194. A successful negotiation occurs when _____.

    a) the interests of both sides are satisfied
    b) both sides concede an equal amount
    c) both sides agree to disagree
    d) the negotiation is settled by a federal arbitrator

195. Confrontation _____.

    a) always leads to conflict
    b) implies arguing and hurt feelings
    c) can relieve stress
    d) should always be done privately

196. Creative thinking _____.

    a) is limited to people who were born with creativity
    b) is limited to completely new ideas and technology
    c) can only be proven through implementation
    d) is easier for younger people

## Personal/Professional Effectiveness

197. Which of the following is NOT true about written communication?

   a) Readers can infer the author's tone from his/her body language.
   b) A key to good writing is rewriting.
   c) The level of words and writing style is dependent on the audience.
   d) Careful selection of words is important for e-mails.

198. An engineering supervisor is preparing an agenda for the monthly engineering department meeting. She realizes that there are too many items to be covered in the usual 2-hour meeting. For an effective meeting, she should _____.

   a) prioritize the topics and save the less urgent topics for the next meeting
   b) limit attendance to only senior engineers
   c) extend the meeting to 3 hours
   d) limit the discussion time on all the items to 2 minutes per item

199. The best format for communicating a project summary to upper management would be a(n) _____.

   a) proposal
   b) letter
   c) report
   d) e-mail

200. The plant manager of a large company is giving an "end-of-year" quality summary to all employees. The style of communication will most likely be _____.

   a) informal with no audience interaction
   b) formal with some audience interaction
   c) informal with audience interaction
   d) formal with no audience interaction

# Solutions

## MATHEMATICAL FUNDAMENTALS

**1. b**

$$\log_{10} x = 4 - \log_{10}(4x-2)$$

$$\log_{10} x + \log_{10}(4x-2) = 4$$

Referencing *Fundamentals of Manufacturing*, 2nd Edition (Rufe 2002), Eq. 1-14 (p. 3),

$$\log_{10} x(4x-2) = 4$$

$$\log_{10}(4x^2 - 2x) = 4$$

$$10^{\log(4x^2 - 2x)} = 10^4$$

$$4x^2 - 2x = 10{,}000$$

$$4x^2 - 2x - 10{,}000 = 0$$

$$\frac{2 \pm \sqrt{(-2)^2 - 4(4)(-10{,}000)}}{2(4)} = 50.25,\ -49.75$$

**2. b**

$$6x + 3y = 10$$
$$3x + 4y = 20$$

$$6x + 3y = 10$$
$$-2[3x + 4y = 20]$$

$$6x + 3y = 10$$
$$-6x - 8y = -40$$

$$-5y = -30$$
$$y = 6$$

$$6x + 3(6) = 10$$
$$6x = -8$$
$$x = -1.33$$

## Solutions

**3. c**

Referencing *Fundamentals of Manufacturing*, 2nd Edition (Rufe 2002), Figure 1-3 (p. 9),

$$V_{cyl} = \pi r^2 h = 10 \text{ m}^3$$

$$r = \frac{d}{2}$$

$$h = 2d$$

$$V_{cyl} = \pi \left(\frac{d}{2}\right)^2 2d$$

$$V_{cyl} = \frac{\pi d^3}{2}$$

$$10 \text{ m}^3 = \frac{\pi d^3}{2}$$

$$d = \sqrt[3]{\frac{2(10 \text{ m}^3)}{\pi}}$$

$$= 1.85 \text{ m}$$

**4. d**

A line parallel to the line in question must have the same slope. The slope of the given line is 1/2. If the possible answers are given in standard form referencing *Fundamentals of Manufacturing*, 2nd Edition (Rufe 2002), Eq. 1-31 (p. 8), ($y = mx + b$), the only line with the same slope is answer "d."

a) $y = -\frac{1}{2}x + 4$, $m = -1/2$

b) $y = 2x + 4$, $m = 2$

c) $y = 2x - 8$, $m = 2$

d) $y = \frac{1}{2}x + 1$, $m = 1/2$

**5. c**

$x_1 = 10, y_1 = 10$

$x_2 = 2, y_2 = 3$

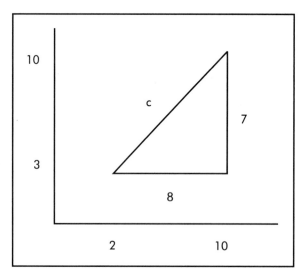

Q5 solution.

Referencing *Fundamentals of Manufacturing*, 2nd Edition (Rufe 2002), Eq. 1-36 (p. 11),

$$c^2 = a^2 + b^2$$

$$a = 7$$

$$b = 8$$

$$c = \sqrt{a^2 + b^2} = \sqrt{8^2 + 7^2} = 10.6$$

**6. d**

Referencing *Fundamentals of Manufacturing*, 2nd Edition (Rufe 2002), Eq. 1-45 (p. 13), the committees possible with Diane:

$$C(n,r) = \frac{n!}{r!(n-r)!}$$

$$n = 5$$
$$r = 3$$
$$C(5,3) = \frac{5!}{3!(5-3)!} = 10$$

Committees possible without Diane:

$$n = 4$$
$$r = 3$$
$$C(4,3) = \frac{4!}{3!(4-3)!} = 4$$

$$C_{with\ Diane} - C_{without\ Diane} = 10 - 4 = 6$$

$$P(\text{Diane on a committee}) = \frac{6}{10} = 60\%$$

**7. c**

Referencing *Fundamentals of Manufacturing*, 2nd Edition (Rufe 2002), Eq. 1-51 (p. 14),

$$\bar{x} = \frac{1}{n}\sum_{i=1}^{n} x_i = \frac{1}{10}(60 + 72 + 65 + \ldots + 80)$$

$$= \frac{700}{10} = 70 \text{ in.}$$

and Eq. 1-56 (p. 14),

$$s = \sqrt{\frac{\sum_{i=1}^{n} x_i^2 - n(\bar{x})^2}{n-1}}$$

$$= \sqrt{\frac{(61^2 + 72^2 + \ldots + 80^2) - 10(70)^2}{10 - 1}} = 5.9 \text{ in.}$$

**8. c**

Referencing *Fundamentals of Manufacturing*, 2nd Edition (Rufe 2002), Eq. 1-57 (p. 16),

$$z_i = \frac{x_i - \mu}{\sigma}$$

$$x_{UCL} = 0.225$$
$$x_{LCL} = 0.217$$
$$\mu = 0.220$$
$$\sigma = 0.002$$

$$z_1 = \frac{x_{UCL} - \mu}{\sigma} = \frac{0.225 - 0.220}{0.002} = 2.5$$

$$z_2 = \frac{x_{LCL} - \mu}{\sigma} = \frac{0.217 - 0.220}{0.002} = -1.5$$

and Table 1-1 (pp. 18–19),

Area to the left of $z_1$ = 0.9938
Area to the left of $z_2$ = 0.0668

The percentage of acceptable parts (parts between the UCL and LCL) is:

Area between $z_1$ and $z_2$ = 0.9938 − 0.0668
= 0.972 × 100
= 92.7%

## Solutions

**9. a**

$f(x) = 2x^3 + 3x^2 + 4$

Referencing *Fundamentals of Manufacturing*, 2nd Edition (Rufe 2002), Eq. 1-61 (p. 20),

$f'(x) = 3(2x^{3-1}) + 2(3x^{2-1}) + 0 = 6x^2 + 6x$
$f''(x) = 2(6x^{2-1}) + 1(6x^{1-0}) = 12x + 6$
$f'(x) = 6x(x + 1) = 0$ when $x = 0, x = -1$
$f''(0) = 12(0) + 6 = 6$
$f''(-1) = 12(-1) + 6 = -6$

relative maximum is at
  $x = -1$ since $f''(-1) < 0$

relative minimum is at
  $x = 0$ since $f''(0) > 0$

**10. b**

$f(x) = 3x^2 + 2x + 1$

Referencing *Fundamentals of Manufacturing*, 2nd Edition (Rufe 2002), Eq. 1-68 (p. 22),

$$\int_1^{10} 3x^2 + 2x + 1$$

$$= \frac{1}{2+1}(3x^{2+1}) + \frac{1}{1+1}(2x^{1+1}) + x$$
$$= x^3 + x^2 + x$$

and Eq. 1-64 (p. 22),

$$f(x^3 + x^2 + x)\Big|_{a=1}^{b=10}$$
$$= (10^3 + 10^2 + 10) - (1^3 + 1^2 + 1) = 1{,}107$$

# PHYSICS AND ENGINEERING SCIENCES

## 11. b

Referencing *Fundamentals of Manufacturing*, 2nd Edition (Rufe 2002), Table D-2 (p. 397),

$$0.00020 \text{ in.} \times \frac{0.0254 \text{ m}}{\text{in.}} \times \frac{1 \times 10^6 \text{ microns}}{\text{m}}$$
$$= 5.08 \text{ microns}$$

## 12. b

Referencing *Fundamentals of Manufacturing*, 2nd Edition (Rufe 2002), Figure 1-3 (p. 9),

$$V_{tank} = \pi r^2 h$$
$$d = 20 \text{ in.}$$
$$r = 10 \text{ in.}$$
$$h = 48 \text{ in.}$$
$$V_{tank} = \pi (10 \text{ in.})^2 (48 \text{ in.}) = 15{,}079.6 \text{ in.}^3$$

and Table 2-7 (p. 32),

$$15{,}079.6 \text{ in.}^3 \times \frac{1 \text{ gal}}{231 \text{ in.}^3} = 65.3 \text{ gal}$$

At the water to coolant concentrate ratio of 3:1, 48 gal of water and 16 gal of coolant concentrate would provide 64 gal total. Since gallons of coolant concentrate can not be split into fractions, the maximum that can be used is 16 gal of concentrate.

## 13. d

According to *Fundamentals of Manufacturing*, 2nd Edition (Rufe 2002), p. 34, *reflected light* is the portion of the light wave redirected away from the object.

## 14. d

Referencing *Fundamentals of Manufacturing*, 2nd Edition (Rufe 2002), Eq. 4-4 (p. 38),

$$\text{Relative intensity} = 10 \log \frac{I}{I_0}$$
$$I = 2 \times 10^{-6} \text{ W/m}^2$$
$$I_o = 1 \times 10^{-12} \text{ W/m}^2$$

Relative intensity =

$$10 \log \frac{2 \times 10^{-6} \text{ W/m}^2}{1 \times 10^{-12} \text{ W/m}^2} = 63 \text{ dB}$$

## 15. b

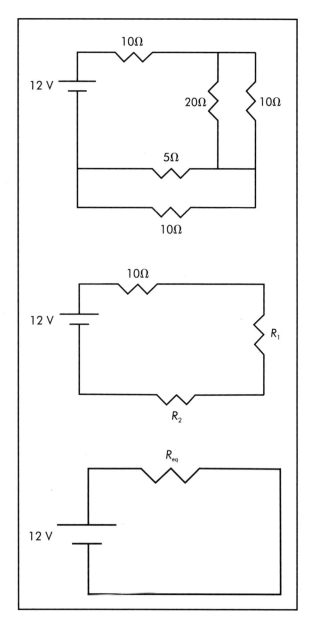

Q15 solution.

Referencing *Fundamentals of Manufacturing*, 2nd Edition (Rufe 2002), Figure 5-3 (p. 44),

$$R_1 = \frac{20(10)}{20+10} = 6.7\Omega$$

$$R_2 = \frac{5(10)}{5+10} = 3.3\Omega$$

$$R_{eq} = 10 + R_1 + R_2$$

$$R_{eq} = 10 + 6.7 + 3.3 = 20\Omega$$

and Eq. 5-4 (p. 42),

$$I = \frac{V}{R} = \frac{12V}{20\Omega} = 0.6 \text{ A}$$

## 16. d

Referencing *Fundamentals of Manufacturing*, 2nd Edition (Rufe 2002), Eq. 5-1 (p. 42),

$$P(\text{sourced}) = IV$$

$$I = 1.5 \text{ A}$$

$$V = 20 \text{ V}$$

$$P(\text{sourced}) = 1.5 \text{ A}(20 \text{ V}) = 30 \text{ W}$$

## 17. c

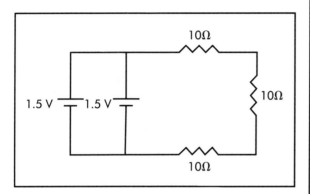

Q17 solution.

Referencing *Fundamentals of Manufacturing*, 2nd Edition (Rufe 2002), Figure 5-3 (p. 44),

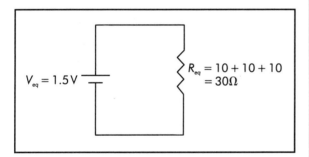

Q17 solution (continued).

and Eq. 5-3 (p. 42),

$$P = \frac{V^2}{R}$$

$$P = \frac{1.5^2}{30} = 0.075 \text{ A} = 75 \text{ mA}$$

## 18. b

Referencing *Fundamentals of Manufacturing*, 2nd Edition (Rufe 2002), Eq. 6-11 (p. 52), and Figure 6-10 (p. 53),

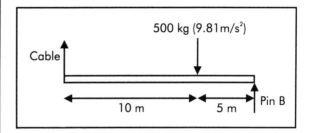

Q18 solution.

$\sum M_{pinB}$
$= -\text{cable}(15 \text{ m}) + 500 \text{ kg}(9.81 \text{m/s}^2)(5 \text{ m}) +$
  $\text{pin B}(0 \text{ m})$
$= 0$

Cable = 1,635 N

### 19. a

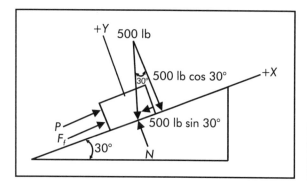

Q19 solution.

Referencing *Fundamentals of Manufacturing*, 2nd Edition (Rufe 2002), Eq. 6-8 (p. 52),

$$\sum F_x = P + F - 500 \sin 30° = 0$$
$$P = -F + 250$$

and Eq. 6-9 (p. 52),

$$\sum F_y = N - 500 \cos 30° = 0$$
$$N = 433 \text{ lb}$$

By Eq. 6-12 (p. 54), $F_f = \mu N$, so

$$P = -\mu N + 250$$
$$P = -0.25(433) + 250$$
$$P = 142 \text{ lb}$$

### 20. b

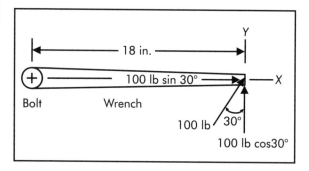

Q20 solution.

Referencing *Fundamentals of Manufacturing*, 2nd Edition (Rufe 2002), Eq. 6-7 (p. 51), *torque* or *moment* is equal to force times its perpendicular distance from the point of rotation. The line of action of the $X$ component of the 100 lb passes through the point of rotation. Therefore, it does not create a moment or torque. The only force creating a moment or torque is the $Y$ component of the 100 lb.

$$\begin{aligned}\text{Torque} &= F \times d \\ &= 100 \text{ lb} \cos 30° \times 18 \text{ in.} \\ &= 1{,}559 \text{ in.-lb}\end{aligned}$$

## 21. d

Referencing *Fundamentals of Manufacturing*, 2nd Edition (Rufe 2002), Eq. 7-4 (p. 57),

$$s = v_0 t + \frac{at^2}{2}$$

$v_0 = 0$

$a = 9.81 \text{ m/s}^2$

$t = 1.5 \text{ s}$

$$s = 0(1.5 \text{ s}) + \frac{9.81 \text{ m/s}^2 (1.5 \text{ s})^2}{2} = 11 \text{ m}$$

## 22. b

Referencing *Fundamentals of Manufacturing*, 2nd Edition (Rufe 2002), Eq. 7-11 (p. 58),

$$\omega = \omega_0 + \alpha t$$

$\omega_0 = 0$

$\alpha = 2 \text{ rad/s}^2$

$$\omega = 200 \frac{\text{rev}}{\text{min}} \times \frac{\text{min}}{60 \text{ s}} \times 2\pi \frac{\text{rad}}{\text{rev}} = 20.9 \text{ rad/s}$$

$$t = \frac{\omega - \omega_0}{\alpha} = \frac{20.9 \text{ rad/s} - 0}{2 \text{ rad/s}^2} = 10.5 \text{ s}$$

and Eq. 7-12 (p. 58),

$$\theta = \omega_0 t + \frac{\alpha t^2}{2}$$

$$\theta = 0(10.5 \text{ s}) + \frac{2 \text{ rad/s}^2 (10.5 \text{ s})^2}{2}$$

$$= 110.3 \text{ rad}$$

$$110.3 \text{ rad} \times \frac{\text{rev}}{2\pi \text{ rad}} = 18 \text{ revolutions}$$

## 23. c

Referencing *Fundamentals of Manufacturing*, 2nd Edition (Rufe 2002), Eq. 8-4 (p. 64),

$$\Delta L = \frac{PL}{AE}$$

$\Delta L = 0.010 \text{ in.}$

$d = 0.5 \text{ in.}$

$E = 30 \times 10^6 \text{ psi}$

$L = 12 \text{ in.}$

$$P = \frac{\Delta L A E}{L}$$

$$= \frac{0.010 \left(\frac{\pi (0.500)^2}{4}\right)(30 \times 10^6)}{12}$$

$$= 4,909 \text{ lb}$$

## 24. b

Referencing *Fundamentals of Manufacturing*, 2nd Edition (Rufe 2002), Eq. 8-2 (p. 63),

$$\sigma = \frac{P}{A}$$

$P = 12 \text{ kN} = 12{,}000 \text{ N}$
$\sigma = 50 \text{ MPa} = 50 \times 10^6 \text{ Pa}$

$$A = \frac{P}{\sigma}$$

$$\frac{\pi d^2}{4} = \frac{P}{\sigma}$$

$$d^2 = \frac{4P}{\pi \sigma}$$

$$d = \sqrt{\frac{4P}{\pi \sigma}}$$

$$= \sqrt{\frac{4(12{,}000)}{\pi(50 \times 10^6)}} = 0.018 \text{ m} = 18 \text{ mm}$$

## 25. d

Referencing *Fundamentals of Manufacturing*, 2nd Edition (Rufe 2002), Eq. 8-10 (p. 67),

$$\theta = \frac{TL}{JG}$$

$T = 5{,}000$ in.-lb
$L = 2$ ft $= 24$ in.
$G = 3.9 \times 10^6$ psi

and Eq. 8-9 (p. 67),

$$J = \pi \left( \frac{d_o^4 - d_i^4}{32} \right)$$

$d_o = 2$ in.
$d_i = 1.5$ in.

$$J = \pi \left( \frac{(2)^4 - (1.5)^4}{32} \right) = 1.07 \text{ in.}^4$$

$$\theta = \frac{5{,}000(24)}{1.07 \text{ in.}^4 (3.9 \times 10^6)} = 0.029 \text{ rad}$$

## 26. c

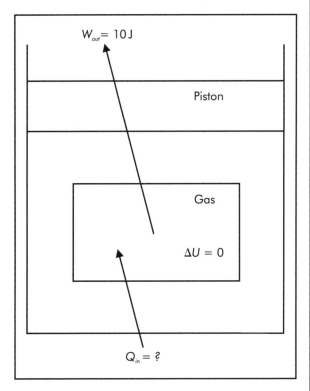

Q26 solution.

Referencing *Fundamentals of Manufacturing*, 2nd Edition (Rufe 2002), Eq. 9-7 (p. 72),

$Q = \Delta U + W = 0 + 10 \text{ J} = 10 \text{ J}$

## 27. a

Referencing *Fundamentals of Manufacturing*, 2nd Edition (Rufe 2002), p. 73, *convection* is the transfer of energy from a surface by the flow of fluid over the surface.

## 28. d

Referencing *Fundamentals of Manufacturing*, 2nd Edition (Rufe 2002), Eq. 9-5 (p. 70),

$\Delta D = \alpha D \Delta T$
$\Delta D = 0.005$ in.

$\alpha = 12 \times 10^{-6} \dfrac{1}{°C}$

$\Delta T = T_i - T_f$

$T_i = 22°$ C

$\Delta T = \dfrac{\Delta D}{\alpha D}$

$= \dfrac{0.005 \text{ in.}}{12 \times 10^{-6} \dfrac{1}{°C} \times 12 \text{ in.}} = 34.7°\text{C}$

$T_f = T_i - \Delta T$
$= 22°\text{ C} - 34.7°\text{ C} = -12.7°\text{ C} \approx -13°\text{ C}$

## 29. d

Referencing *Fundamentals of Manufacturing*, 2nd Edition (Rufe 2002), Eq. 10-11 (p. 78),

$$d_1^2 v_1 = d_2^2 v_2$$

$$d_1 = 1 \text{ in.}$$

Flow rate $= 5 \dfrac{\text{gal}}{\text{min}}$

$$v_1 = \dfrac{\text{Flow rate}}{A_1}$$

$$A_1 = \dfrac{\pi d_1^2}{4}$$

and from Table 2-7 (p. 32), 1 gal = 231 in.³,

$$v_1 = \dfrac{5\dfrac{\text{gal}}{\text{min}} \times 231 \dfrac{\text{in.}^3}{\text{gal}}}{\dfrac{\pi (1 \text{ in.})^2}{4}} = 1{,}471 \text{ psi}$$

$$d_2 = 0.5 \text{ in.}$$

$$v_2 = \dfrac{d_1^2 v_1}{d_2^2}$$

$$= \dfrac{(1 \text{ in.})^2 \left(1{,}471 \dfrac{\text{in.}}{\text{min}}\right)}{(0.5 \text{ in.})^2}$$

$$= 5{,}882 \text{ in./min}$$

## 30. c

Referencing *Fundamentals of Manufacturing*, 2nd Edition (Rufe 2002), Eq. 10-8 (p. 76),

$$P = \dfrac{F}{A}$$

$$P = 100 \text{ psi}$$

$$d = 3 \text{ in.}^2$$

$$F = AP = \dfrac{\pi d^2}{4}(P)$$

$$= \left(\dfrac{\pi (3 \text{ in.})^2}{4}\right) 100 \text{ psi} = 707 \text{ lb}$$

## MATERIALS

### 31. b

Referencing *Fundamentals of Manufacturing*, 2nd Edition (Rufe 2002), Figure 12-9 (p. 98), the minimum quench temperature for forming martensite in plain carbon steel is approximately 220° C. It is the $M_S$ line.

### 32. c

Referencing *Fundamentals of Manufacturing*, 2nd Edition (Rufe 2002), Figure 12-10 (p. 101), the minimum temperature is approximately 800° C to fully anneal 1040 steel (0.40% carbon).

### 33. d

Referencing *Fundamentals of Manufacturing*, 2nd Edition (Rufe 2002), p. 104, *cold working* is another name for *strain hardening*. As a metal is deformed below its recrystallization temperature, it becomes harder.

### 34. c

Referencing *Fundamentals of Manufacturing*, 2nd Edition (Rufe 2002), p. 88, percentage of elongation is a measure of ductility. The higher the percentage, the more the material deformed prior to failure during the tensile test. Therefore, higher percentage of elongation values indicate higher ductility.

### 35. d

According to *Fundamentals of Manufacturing*, 2nd Edition (Rufe 2002), Table 12-2 (p. 102), molybdenum increases creep resistance at elevated temperatures.

### 36. a

From *Fundamentals of Manufacturing*, 2nd Edition (Rufe 2002), Table 13-1 (p. 110), phenolic is the most dense (highest specific gravity).

### 37. d

Referencing *Fundamentals of Manufacturing*, 2nd Edition (Rufe 2002), p. 109, crystallinity increases density and makes the plastic opaque. It also increases strength.

### 38. c

From *Fundamentals of Manufacturing*, 2nd Edition (Rufe 2002), pp. 109 and 113, phenolic is not a thermoplastic.

### 39. c

According to *Fundamentals of Manufacturing*, 2nd Edition (Rufe 2002), pp. 116–117, in a composite material much of the overall strength comes from the fibers rather than the matrix.

## 40. d

From *Fundamentals of Manufacturing*, 2nd Edition (Rufe 2002), p. 119, ceramics have high temperature resistance and corrosion resistance.

## PRODUCT DESIGN

### 41. d

According to *Fundamentals of Manufacturing*, 2nd Edition (Rufe 2002), p. 131, unilateral tolerances allow variation in only one direction from the nominal.

### 42. c

From *Fundamentals of Manufacturing*, 2nd Edition (Rufe 2002), p. 130, the maximum material condition of the hole is the lower limit, 0.620 in.

### 43. d

From *Fundamentals of Manufacturing*, 2nd Edition (Rufe 2002), p. 130, tighter (smaller) tolerances lead to higher manufacturing costs.

### 44. a

From *Fundamentals of Manufacturing*, 2nd Edition (Rufe 2002), p. 130, the allowance is the minimum clearance between mating parts.

Smallest bore diameter = 0.996 in.
Largest shaft diameter = 0.992 in.

Allowance = smallest bore diameter − largest shaft diameter

= 0.996 − 0.992 = 0.004 in.

### 45. b

The largest clearance is when the hole is at LMC and the shaft is at LMC. From the RC6 table in *Machinery's Handbook*, 23rd edition (units are in thousandths of an inch) (Oberg et al. 1988):

| Limits of Clearance | Hole | Shaft |
|---|---|---|
| 1.6 | 2.0 | −1.6 |
| 4.8 | 0 | −2.8 |

LMC hole = 1.000 + 0.0020 = 1.0020
LMC shaft = 1.000 − 0.0028 = 0.9972

Maximum clearance = LMC hole − LMC shaft
= 1.0020 − 0.9972
= 0.0048 in.

### 46. a

From *Fundamentals of Manufacturing*, 2nd Edition (Rufe 2002), p. 133, the part with the smoothest surface has the smallest average roughness, 10 μin.

## 47. d

$$-A + B - C - X = 0$$

$$A = \frac{1}{2}(10.0) = 5 \text{ in.}$$

$$B = 0.1 + 10 + 0.1 = 10.2 \text{ in.}$$

$$C = \frac{1}{2}(3.4) = 1.7 \text{ in.}$$

$$-5 + 10.2 - 1.7 - X = 0$$

$$X = 3.5 \text{ in.}$$

## 48. a

According to *Fundamentals of Manufacturing*, 2nd Edition (Rufe 2002), p. 137, *Rule 1* states that where only a tolerance of size is specified, the limits of size for an individual feature prescribe the extent to which variations in the feature's form as well as its size, are allowed.

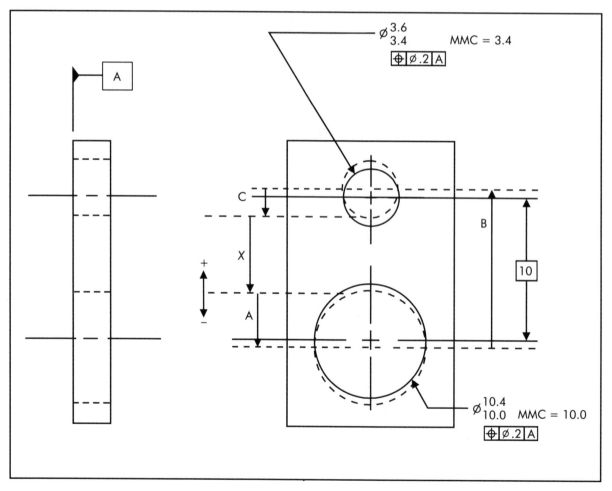

Q47 solution.

**49. d**

According to *Fundamentals of Manufacturing*, 2nd Edition (Rufe 2002), p. 139, and *TMEH Desk Edition* (Cubberly and Bakerjian 1988), p. 9-1, the *feature control frame* states the requirements or instructions for the features to which it is attached. It specifies the type, size and shape of the geometric tolerance zone, dictates datums, and assigns material condition modifiers.

**50. a**

From *Fundamentals of Manufacturing*, 2nd Edition (Rufe 2002), p. 139, the symbol ∅ before a tolerance value indicates the tolerance zone it applies to is cylindrical.

**51. c**

Based on *Fundamentals of Manufacturing*, 2nd Edition (Rufe 2002), Figure 17-4 (p. 141), the flatness control on surface A limits the flatness error to 0.2 mm.

**52. a**

From *Fundamentals of Manufacturing*, 2nd Edition (Rufe 2002), p. 140 and Figure 17-6 (p. 141), the 0.3 tolerance only applies when the hole is at its MMC (10.0). Therefore, the pin diameter required to check straightness is 10.0 – 0.3 = 9.7 mm.

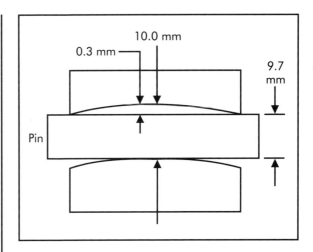

*Q52 solution.*

**53. b**

From *Fundamentals of Manufacturing*, 2nd Edition (Rufe 2002), Figure 17-11 (p. 144), the perpendicularity control applies to the part's rectangular surface, which indicates the tolerance zone is two parallel planes.

**54. c**

From *Fundamentals of Manufacturing*, 2nd Edition (Rufe 2002), p. 144, *runout* is a combination of geometric tolerances used to control the relationship of one or more features to a datum axis.

## Solutions

**55. d**

From *Fundamentals of Manufacturing*, 2nd Edition (Rufe 2002), p. 154, and the *TMEH Desk Edition* (Cubberly and Bakerjian 1988), pp. 3–7, a *solid model* allows for the rapid calculation and extraction of physical property information such as volume, area, centroid, moment of inertia, etc.

**56. a**

*Fundamentals of Manufacturing*, 2nd Edition (Rufe 2002), pp. 154–155, discusses some of the advantages and disadvantages of rapid prototyping.

**57. d**

The process of "picking points" off of a part surface to generate a 3D-CAD model is known as *reverse engineering*. Based on the information in Chapter 18 of *Fundamentals of Manufacturing*, 2nd Edition (Rufe 2002), several of the incorrect choices can be eliminated.

**58. a**

According to Chapter 18 of *Fundamentals of Manufacturing*, 2nd Edition (Rufe 2002), pp. 153–154, the "STL" file is the only correct choice. STL files are commonly used for rapid prototyping.

**59. a**

From *Fundamentals of Manufacturing*, 2nd Edition (Rufe 2002), p. 159, two components can be produced as a single component if they do not move relative to each other.

**60. b**

From *Fundamentals of Manufacturing*, 2nd Edition (Rufe 2002), pp. 159–160, vibration from abnormal tire wear would be considered a failure effect. Vibration is the effect perceived by the customer.

**61. b**

From *Fundamentals of Manufacturing*, 2nd Edition (Rufe 2002), p. 160, QFD is driven by the voice of the customer.

**62. c**

From *Fundamentals of Manufacturing*, 2nd Edition (Rufe 2002), pp. 162–165, group technology can reduce part proliferation and redundant designs.

**63. b**

From *Fundamentals of Manufacturing*, 2nd Edition (Rufe 2002), p. 166, with respect to group technology, attribute coding requires the same number of digits in the code as there are unique part attributes.

**64. d**

From *Fundamentals of Manufacturing*, 2nd Edition (Rufe 2002), p. 160, high RPN values signal problems.

**65. d**

From *Fundamentals of Manufacturing*, 2nd Edition (Rufe 2002), p. 159, *design efficiency* measures how good a product design is with respect to assembly.

# MANUFACTURING PROCESSES

## 66. a

Referencing *Fundamentals of Manufacturing*, 2nd Edition (Rufe 2002), Eq. 20-1 (p. 173),

$$C = VT^n$$

$$C = 200 \frac{\text{ft}}{\text{min}}$$

$$V = 100 \frac{\text{ft}}{\text{min}}$$

$$n = 0.125$$

$$200 = 100 T^{0.125}$$

$$\frac{200}{100} = T^{0.125}$$

$$\ln\frac{200}{100} = \ln T^{0.125}$$

$$\ln 2 = 0.125 \ln T$$

$$\frac{\ln 2}{0.125} = \ln T$$

$$5.545 = \ln T$$

$$e^{5.545} = e^{\ln T}$$

$$T = e^{5.545} = 256 \text{ min}$$

## 67. d

From *Fundamentals of Manufacturing*, 2nd Edition (Rufe 2002), Table 20-3 (p. 175), diamond has the highest wear resistance combined with poor impact strength.

## 68. d

$$D = 4 \text{ in.}$$
$$d = 0.060 \text{ in.}$$
$$V_c = 100 \frac{\text{ft}}{\text{min}}$$
$$P_m = 2 \text{ hp}$$
$$E = 80\%$$
$$U = 1$$

Referencing *Fundamentals of Manufacturing*, 2nd Edition (Rufe 2002), Eq. 21-3 (p. 180),

$$Q = 12 \times d \times f_r \times V_c$$

$$f_r = \frac{Q}{12 d V_c}$$

Eq. 21-6 (p. 180),

$$U = \frac{P_s}{Q}$$

$$Q = \frac{P_s}{U}$$

and Eq. 21-5 (p. 180),

$$P_m = \frac{P_s}{E}$$

$$P_s = E P_m$$

$$Q = \frac{P_s}{U} = \frac{E P_m}{U} = \frac{0.80 \times 2}{1} = 1.6 \frac{\text{in.}^3}{\text{min}}$$

$$f_r = \frac{1.6 \frac{\text{in.}^3}{\text{min}}}{12 \times 0.060 \text{ in.} \times 100 \frac{\text{ft}}{\text{min}}}$$

$$= 0.022 \frac{\text{in.}}{\text{rev}}$$

## Solutions

### 69. b

From *Fundamentals of Manufacturing*, 2nd Edition (Rufe 2002), p. 180, collets provide maximum accuracy but are limited to round workpieces.

### 70. c

$$V_c = 100 \frac{\text{ft}}{\text{min}}$$
$$D = 1 \text{ in.}$$

Referencing *Fundamentals of Manufacturing*, 2nd Edition (Rufe 2002), Eq. 21-7 (p. 182),

$$N = \frac{V_c \times 12}{\pi \times D}$$

$$N = \frac{100 \frac{\text{ft}}{\text{min}} \times 12 \frac{\text{in.}}{\text{ft}}}{(\pi \times 1) \frac{\text{in.}}{\text{rev}}} = 382 \text{ rpm}$$

### 71. a

$$n = 4 \text{ teeth}$$
$$V_c = 100 \frac{\text{ft}}{\text{min}}$$
$$F = 4 \frac{\text{in.}}{\text{min}}$$
$$P_m = 3 \text{ hp}$$
$$E = 75\%$$
$$w = 1 \text{ in.}$$
$$U = 1$$

Referencing *Fundamentals of Manufacturing*, 2nd Edition (Rufe 2002), Eq. 21-15 (p.184),

$$Q = w \times d \times f_t \times n \times N$$
$$d = \frac{Q}{w \times f_t \times n \times N}$$

Eq. 21-17 (p. 185),

$$U = \frac{P_s}{Q}$$
$$Q = \frac{P_s}{U}$$

Eq. 21-16 (p. 184),

$$P_m = \frac{P_s}{E}$$
$$P_s = E P_m$$
$$Q = \frac{E(P_m)}{U}$$
$$= \frac{0.75(3)}{1} = 2.25 \frac{\text{in.}^3}{\text{min}}$$

Eq. 21-12 (p. 184),

$$N = \frac{V_c \times 12}{\pi \times D}$$
$$= \frac{100 \frac{\text{ft}}{\text{min}} \times 12 \frac{\text{in.}}{\text{ft}}}{(\pi \times 1) \frac{\text{in.}}{\text{rev}}} = 382 \text{ rpm}$$

and Eq. 21-14 (p. 184),

$$F = f_t \times n \times N$$
$$f_t = \frac{F}{n \times N}$$
$$= \frac{4 \frac{\text{in.}}{\text{min}}}{4 \text{ teeth} \times 382 \text{ rpm}} = 0.0026 \frac{\text{in.}}{\text{tooth}}$$

$$d = \frac{2.25 \frac{\text{in.}^3}{\text{min}}}{1 \text{ in.} \times 0.0026 \frac{\text{in.}}{\text{tooth}} \times 4 \text{ teeth} \times 382 \text{ rpm}}$$
$$= 0.566 \text{ in.}$$

## 72. d

From *Fundamentals of Manufacturing*, 2nd Edition (Rufe 2002), p. 184, climb milling generally requires less feed force than up milling.

## 73. a

From *Fundamentals of Manufacturing*, 2nd Edition (Rufe 2002), p. 186, the number of teeth per inch is called the *pitch*.

## 74. c

According to *Fundamentals of Manufacturing*, 2nd Edition (Rufe 2002), Table 22-1 (p. 194), cold formed parts have good surface finishes.

## 75. d

From *Fundamentals of Manufacturing*, 2nd Edition (Rufe 2002), p. 194, forging aligns the grain to the contour of the part, thus increasing its strength.

## 76. d

From *Fundamentals of Manufacturing*, 2nd Edition (Rufe 2002), p. 194, coining can produce great detail with only a small amount of deformation.

## 77. d

$F$ = 40 tons
$S$ = 20,000 psi
$L$ = 2 in. + 2 in. + 2 in. + 2 in. = 8 in.

According to *Fundamentals of Manufacturing*, 2nd Edition (Rufe 2002), Eq. 23-1 (p. 199),

$F = SLT$

$$T = \frac{F}{SL} = \frac{40 \text{ ton} \times 2{,}000 \frac{\text{lb}}{\text{ton}}}{20{,}000 \text{ psi} \times 8 \text{ in.}}$$
$= 0.500$ in.

## 78. d

From *Fundamentals of Manufacturing*, 2nd Edition (Rufe 2002), p. 202, completely folding over the edge of a piece of sheet metal to provide strength and rigidity, such as on the edges of an automobile door, is known as *hemming*.

## 79. b

From *Fundamentals of Manufacturing*, 2nd Edition (Rufe 2002), p. 200, lancing produces holes without completely removing the metal.

## 80. d

From *Fundamentals of Manufacturing*, 2nd Edition (Rufe 2002), p. 206, binders increase part strength after compaction (prior to sintering).

## 81. b

From *Fundamentals of Manufacturing*, 2nd Edition (Rufe 2002), p. 207, the second stage (high temperature) of the sintering cycle fuses the particles together.

## 82. b

From *Fundamentals of Manufacturing*, 2nd Edition (Rufe 2002), p. 210, the parts of the gating system that lead directly into the mold cavity are the gates.

## 83. d

From *Fundamentals of Manufacturing*, 2nd Edition (Rufe 2002), p. 213, *investment casting* provides superior dimensional accuracy and surface finish.

## 84. a

From *Fundamentals of Manufacturing*, 2nd Edition (Rufe 2002), p. 218, stick-electrode welding forms a protective slag layer over the weld.

## 85. c

From *Fundamentals of Manufacturing*, 2nd Edition (Rufe 2002), p. 220, *resistance spot welding* does not require filler metal or flux.

## 86. d

From *Fundamentals of Manufacturing*, 2nd Edition (Rufe 2002), p. 221, ultrasonic welding uses high-frequency vibration and pressure to join metals.

## 87. d

From *Fundamentals of Manufacturing*, 2nd Edition (Rufe 2002), p. 225, the electric charge applied to the paint particles in electrostatic spraying increases the transfer efficiency due to the magnetic attraction.

## 88. b

From *Fundamentals of Manufacturing*, 2nd Edition (Rufe 2002), p. 225, electroplating requires a DC power source.

## 89. a

From *Fundamentals of Manufacturing*, 2nd Edition (Rufe 2002), p. 230, *blow molding* is used to produce bottle-shaped products. A hollow plastic gas tank is a similar product.

## 90. d

From *Fundamentals of Manufacturing*, 2nd Edition (Rufe 2002), p. 227, *extrusion* is used to produce plastic parts, such as plastic pipe or tubing.

**91. c**

From *Fundamentals of Manufacturing*, 2nd Edition (Rufe 2002), p. 238, *pultrusion* produces parts with a constant cross-section from a composite material.

**92. b**

From *Fundamentals of Manufacturing*, 2nd Edition (Rufe 2002), p. 243, ceramic parts used in engineering applications, such as spark plugs, are generally made from crystalline ceramics by *isostatic pressing*.

**93. c**

From *Fundamentals of Manufacturing*, 2nd Edition (Rufe 2002), p. 242, the part's wall thickness in slip casting is determined by the amount of time the slip is in the plaster mold. The plaster mold extracts water from the liquid slip leaving solid clay. The more time the slip is left in the plaster mold, the more water extracted, thereby increasing the wall thickness.

**94. d**

From *Fundamentals of Manufacturing*, 2nd Edition (Rufe 2002), p. 245, the process of making a printed circuit board is called *selective etching*.

**95. d**

From *Fundamentals of Manufacturing*, 2nd Edition (Rufe 2002), p. 247, surface-mounted devices are temporarily secured by solder paste before soldering.

## PRODUCTION SYSTEMS

### 96. b

From *Fundamentals of Manufacturing*, 2nd Edition (Rufe 2002), p. 251, the *assemble-to-order environment* begins the assembly of a product after the customer's order. Products are made from common components, which are possibly stocked.

### 97. c

From *Fundamentals of Manufacturing*, 2nd Edition (Rufe 2002), p. 252, *forecasting* becomes more accurate when the forecast period is short.

### 98. b

From *Fundamentals of Manufacturing*, 2nd Edition (Rufe 2002), p. 253, the master production schedule does not determine where bottlenecks will occur in the manufacturing process.

### 99. d

From *Fundamentals of Manufacturing*, 2nd Edition (Rufe 2002), Figure 32-2 (p. 256), personnel data is not an element of material requirements planning.

### 100. b

From *Fundamentals of Manufacturing*, 2nd Edition (Rufe 2002), p. 253, *capacity requirements planning* considers work-in-process and replacement parts.

### 101. d

From *Fundamentals of Manufacturing*, 2nd Edition (Rufe 2002), p. 255, the *critical ratio* scheduling technique is a function of due date and lead time.

### 102. c

From p. 255, *MRP* is a time-phased material planning system.

### 103. c

From *Fundamentals of Manufacturing*, 2nd Edition (Rufe 2002), p. 257, *MRPII* plans and schedules all the resources in a manufacturing company.

### 104. a

From *Fundamentals of Manufacturing*, 2nd Edition (Rufe 2002), p. 263, a *pull system* only produces parts as required by the subsequent operation.

## 105. c

From *Fundamentals of Manufacturing*, 2nd Edition (Rufe 2002), p. 260, lean production utilizes multi-skilled workers and flexible equipment to produce a variety of cost-competitive products.

## 106. d

From *Fundamentals of Manufacturing*, 2nd Edition (Rufe 2002), p. 261, a *kanban* signals the need for stock replenishment.

## 107. d

From *Fundamentals of Manufacturing*, 2nd Edition (Rufe 2002), p. 260, lean inventory levels are continually reduced.

## 108. b

From *Fundamentals of Manufacturing*, 2nd Edition (Rufe 2002), p. 261, call lights or andon boards are examples of *visual control*.

## 109. a

From *Fundamentals of Manufacturing*, 2nd Edition (Rufe 2002), p. 262, *production leveling* eliminates waste caused by overtime and fatigue while ensuring a constant flow from suppliers and the proper utilization of workers.

## 110. b

From *Fundamentals of Manufacturing*, 2nd Edition (Rufe 2002), p. 261, installing a vision system to detect missing parts on an assembly prior to shipping is an example of *mistake proofing*.

## 111. c

Referencing *Fundamentals of Manufacturing*, 2nd Edition (Rufe 2002), Eq. 33-1 (p. 260),

$$T_t = \frac{A_t}{D_r}$$

$$A_t = 8 \text{ hr} \times 60 \frac{\text{min}}{\text{hr}} - 30 \text{ min} - 15 \text{ min}$$
$$- 15 \text{ min}$$
$$= 420 \text{ min}$$

$$D_r = 600 \frac{\text{cases}}{\text{day}} \times 10 \frac{\text{parts}}{\text{case}} = 6,000 \frac{\text{parts}}{\text{day}}$$

$$T_t = \frac{420 \text{ min}}{6,000 \text{ parts}} = 0.07 \frac{\text{min}}{\text{part}}$$

## 112. d

From *Fundamentals of Manufacturing*, 2nd Edition (Rufe 2002), p. 260, *value stream analysis* identifies value-added and non-value-added steps.

## 113. c

From *Fundamentals of Manufacturing*, 2nd Edition (Rufe 2002), p. 262, the *5S* strategy covers the basic principles of industrial housekeeping.

## 114. d

From *Fundamentals of Manufacturing*, 2nd Edition (Rufe 2002), p. 262, all are true with the exception of "d." The workers' responsibility for quality is not reduced.

## 115. b

From *Fundamentals of Manufacturing*, 2nd Edition (Rufe 2002), p. 265, the *process plan* determines the manufacturing processes used, specifies tooling requirements, and includes routing and operations sheets, etc.

## 116. a

From *Fundamentals of Manufacturing*, 2nd Edition (Rufe 2002), p. 268, the *process layout* groups together similar functions.

## 117. c

$A_t = 60$ sec

$P_r = 75\%$

$P_a = 5\%$

Referencing *Fundamentals of Manufacturing*, 2nd Edition (Rufe 2002), Eq. 34-1 (p. 271),

$$N_t = \frac{A_t \times P_r}{100} = \frac{60 \text{ sec} \times 75}{100} = 45 \text{ sec}$$

and Eq. 34-2 (p. 271),

$$S_t = \frac{N_t(100 + P_a)}{100} = \frac{45 \text{ sec} \times (100 + 5)}{100}$$
$$= 47.3 \text{ sec}$$

## 118. b

From *Fundamentals of Manufacturing*, 2nd Edition (Rufe 2002), p. 267, *nonsynchronous assembly* is used when the time required to perform different operations varies greatly.

## 119. b

From *Fundamentals of Manufacturing*, 2nd Edition (Rufe 2002), p. 266, *jigs* differ from fixtures in that jigs guide the cutting tool and fixtures do not.

## 120. d

From the *Fundamentals of Manufacturing Supplement* (Rufe 2005), p. 94, an *indexing fixture* can be used to machine equally spaced holes for a bolt circle.

### 121. d

From *Fundamentals of Manufacturing*, 2nd Edition (Rufe 2002), p. 266, *CAPP* (computer-aided process planning) is an expert computer system that can specify the machinery to be used for production, the sequence of operations, and tooling for a new part.

### 122. b

From *Fundamentals of Manufacturing*, 2nd Edition (Rufe 2002), p. 266, using the 3-2-1 principle, 6 points are needed to restrict movement with three degrees of freedom.

### 123. a

From *Fundamentals of Manufacturing*, 2nd Edition (Rufe 2002), p. 271, monitoring press conditions and comparing them to a nominal value is an example of *predictive maintenance*. A machine's failure could be predicted if a measured value was significantly different than the nominal value.

### 124. b

From *Fundamentals of Manufacturing*, 2nd Edition (Rufe 2002), p. 268, in a *product-process layout*, different product families are arranged discretely in a plant.

### 125. a

From *Fundamentals of Manufacturing*, 2nd Edition (Rufe 2002), pp. 266–267, a diamond pin locator is an example of a *relieved locator* (three equal flats on the length of the pin). The *Fundamentals of Manufacturing Supplement* (Rufe 2005), p. 88 also discusses diamond pin locators.

### 126. a

From *Fundamentals of Manufacturing*, 2nd Edition (Rufe 2002), p. 267, clamps should be located over a locator if possible.

### 127. c

From *Fundamentals of Manufacturing*, 2nd Edition (Rufe 2002), p. 274, *dependent demands* are derived from the demands for other items.

### 128. b

From *Fundamentals of Manufacturing*, 2nd Edition (Rufe 2002), p. 275, about 80% of the items in inventory will be involved in 20% of the usage measured in dollars.

## 129. c

Referencing *Fundamentals of Manufacturing*, 2nd Edition (Rufe 2002), Eq. 35-1 (p. 274),

$$I_t = \frac{A_y}{A_i}$$

$$A_y = 10{,}000 \frac{\text{dollars}}{\text{month}} \times 12 \text{ months}$$
$$= \$120{,}000$$

$$A_i = \$5{,}000$$

$$I_y = \frac{\$120{,}000}{\$5{,}000} = 24 \text{ turns}$$

## 130. c

From *Fundamentals of Manufacturing*, 2nd Edition (Rufe 2002), p. 274, a low number of inventory turns indicates a supplier carries a large inventory and replenishes it infrequently.

## 131. c

Referencing *Fundamentals of Manufacturing*, 2nd Edition (Rufe 2002), Eq. 35-2 (p. 275),

$$EOQ = \sqrt{\frac{2AS}{ic}}$$

$A = 100{,}000$ units
$S = \$25$
$i = 8\%$
$c = \$10$

$$EOQ = \sqrt{\frac{2 \times 100{,}000 \times 25}{0.08 \times 10}} = 2{,}500$$

## 132. d

From *Fundamentals of Manufacturing*, 2nd Edition (Rufe 2002), p. 275–276, APICS (American Production and Inventory Control Society) is a professional society specializing in inventory management.

## 133. a

From *Fundamentals of Manufacturing*, 2nd Edition (Rufe 2002), pp. 276–277, coordinating the flow of materials, information, and money between customers and suppliers is known as *supply chain management*.

## 134. c

Referencing *Fundamentals of Manufacturing*, 2nd Edition (Rufe 2002), p. 275, machine screws would most likely be considered "C" items since they are typically large in quantity and low in cost.

## 135. c

Referencing *Fundamentals of Manufacturing*, 2nd Edition (Rufe 2002), p. 276, near-zero inventory or just-in-time can be achieved in part by working with the supplier to produce *zero defects*.

## AUTOMATED SYSTEMS AND CONTROL

### 136. b

Referencing *Fundamentals of Manufacturing*, 2nd Edition (Rufe 2002), p. 281, justification for the implementation of automation should include how it will improve manufacturing productivity.

### 137. d

Referencing *Fundamentals of Manufacturing*, 2nd Edition (Rufe 2002), p. 288, the network operating system connects the user with the network.

### 138. c

From *Fundamentals of Manufacturing*, 2nd Edition (Rufe 2002), p. 290, a *protocol* is the predefined manner or set of rules by which a function or service is provided. A protocol regulates the format for moving data.

### 139. d

Referencing *Fundamentals of Manufacturing*, 2nd Edition (Rufe 2002), p. 295, linear and circular interpolation would be required to produce the "B."

### 140. d

*Fundamentals of Manufacturing*, 2nd Edition (Rufe 2002), Figure 38-5 (p. 296), Table B-2 (p. 380), and Table B-3 (p. 381) will help interpret the program.

N10 G00 position (rapid traverse) to X = 0, Y = 0, T4 = load tool #4, M06 = tool change

N20 G00 position (rapid traverse) to X = 1, Y = 1

N30 G01 linear cutting, Z = −.1 depth of cut, F160 = feed rate 16.0 in./min, S4000 = 4,000 rpm spindle speed

N40 G01 linear cutting to X = 1, Y = 3

N50 G02 arc interpolation clockwise (from last position X = 1, Y = 3) to X = 1, Y = 2, centered at I = 1, J = 2.5 (X = 1, Y = 2.5)

N60 G01 linear cutting to X = 2, Y = 1

N70 G00 position (rapid traverse) to Z = 2 (2 in. from the work surface)

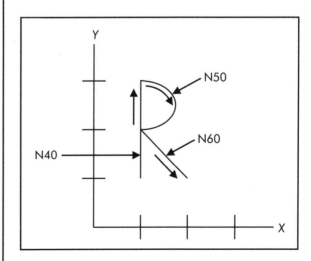

*Q140 solution.*

## Solutions

**141. d**

*Fundamentals of Manufacturing*, 2nd Edition (Rufe 2002), p. 298 describes various control systems. The thermostat provides feedback to the furnace, thereby making it a *closed-loop system*.

**142. b**

Depressing the push-button switch in rung one and holding it in the depressed position energizes output A1. When output A1 energizes, contacts A1-1 in Rung 2 close. Although contacts A1-1 are closed, contacts B1-1 are open. Therefore, Rung 2 will not energize C1.

Contacts D1-1 in Rung 3 are normally closed. Therefore, C1 should be energized. However, in Rung 4, contacts A1-1 are closed (as a result of the push-button switch energizing output A1 in Rung 1), thereby energizing relay D1 and opening contacts D1-1. As a result, Rung 3 does not energize C1. Output C1 remains off.

**143. b**

From *Fundamentals of Manufacturing*, 2nd Edition (Rufe 2002), p. 306, *continuous path control* requires multiple axes controlled simultaneously.

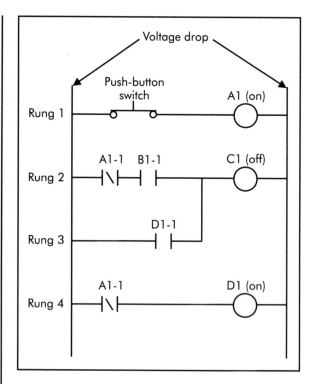

Q142 solution.

**144. a**

From *Fundamentals of Manufacturing*, 2nd Edition (Rufe 2002), p. 306, the *part coordinate system* allows the dispenser to move according to the geometry of the part.

**145. d**

From *Fundamentals of Manufacturing*, 2nd Edition (Rufe 2002), p. 312, the *CCD camera* is commonly used in machine vision.

## QUALITY

### 146. a

From *Fundamentals of Manufacturing*, 2nd Edition (Rufe 2002), p. 315, *TQM* focuses on preventing defects, not detecting them.

### 147. b

From *Fundamentals of Manufacturing*, 2nd Edition (Rufe 2002), p. 316, a recall of Firestone's Wilderness® AT tires on the Ford Explorer® is an example of an *external quality cost*.

### 148. d

From *Fundamentals of Manufacturing*, 2nd Edition (Rufe 2002), p. 317, supplier ISO 9000 registration verifies that the supplier has a quality system in place.

### 149. c

From *Fundamentals of Manufacturing*, 2nd Edition (Rufe 2002), p. 315, a company-wide focus on providing quality service is known as *total quality management*.

### 150. b

Referencing *Fundamentals of Manufacturing*, 2nd Edition (Rufe 2002), p. 317, the Baldrige Award has a few unique criteria, which are not required for ISO 9000 registration. Therefore, ISO 9000 registration does not automatically qualify a company for the Baldrige Award.

### 151. a

From *Fundamentals of Manufacturing*, 2nd Edition (Rufe 2002), p. 319, torque is a quantitative measurement taken by an instrument, which can produce a variable result.

### 152. b

From *Fundamentals of Manufacturing*, 2nd Edition (Rufe 2002), Table 43-2 (p. 323), charting the number of voids (defects) per unit (casting) can be done with a $c$ chart. Chapter 54 of the *Fundamentals of Manufacturing Supplement* (Rufe 2005) also discusses $c$ charts.

### 153. c

Referencing *Fundamentals of Manufacturing*, 2nd Edition (Rufe 2002), p. 320, cutting tool chipping would create *assignable variability*. The part variation can be traced to a specific cause.

## 154. b

From *Fundamentals of Manufacturing*, 2nd Edition (Rufe 2002), p. 319, inspection is a non-value-added activity.

## 155. d

| Subgroup | A | B | C | D | $\bar{x}$ | R |
|---|---|---|---|---|---|---|
| 1 | 25 | 30 | 23 | 28 | 26.50 | 7 |
| 2 | 23 | 24 | 25 | 22 | 23.50 | 3 |
| 3 | 31 | 29 | 25 | 25 | 27.50 | 6 |
| 4 | 25 | 26 | 27 | 25 | 25.75 | 2 |
| 5 | 30 | 29 | 25 | 27 | 27.75 | 5 |

$$\bar{\bar{x}} = 26.20, \ \bar{R} = 4.60$$

Referencing *Fundamentals of Manufacturing*, 2nd Edition (Rufe 2002), Eq. 43-6 (p. 325),

$$UCL_{\bar{x}} = \bar{\bar{x}} + A_2 \bar{R}$$

and Table 43-4 (p. 326), with a subgroup size of 4, $A_2 = 0.729$,

$$UCL_{\bar{x}} = \bar{\bar{x}} + A_2 \bar{R} = 26.20 + 0.729(4.60)$$
$$= 29.6$$

## 156. a

Referencing *Fundamentals of Manufacturing*, 2nd Edition (Rufe 2002), Eq. 43-11 (p. 326),

$$LCL_R = D_3 \bar{R}$$

and Table 43-4 (p. 326), with a subgroup size of 4, $D_3 = 0$,

$$LCL_R = D_3 \bar{R} = 0(4.60) = 0$$

## 157. b

| Subgroup | A | B | C | R |
|---|---|---|---|---|
| 1 | 20 | 19 | 20 | 1 |
| 2 | 21 | 20 | 20 | 1 |
| 3 | 20 | 19 | 19 | 1 |
| 4 | 21 | 20 | 20 | 1 |

$$\bar{R} = 1$$

Referencing *Fundamentals of Manufacturing*, 2nd Edition (Rufe 2002), Eq. 43-12 (p. 327),

$$C_p = \frac{USL - LSL}{6\hat{\sigma}}$$

Eq. 43-13 (p. 327),

$$\hat{\sigma} = \frac{\bar{R}}{d_2}$$

$USL = 20 + 2 = 22$
$LSL = 20 - 2 = 18$

and Table 43-4 (p. 326), with a subgroup size of 3, $d_2 = 1.693$,

$$\hat{\sigma} = \frac{\bar{R}}{d_2} = \frac{1}{1.693} = 0.59$$

$$C_p = \frac{UCL - LCL}{6\hat{\sigma}} = \frac{22 - 18}{6(0.59)} = 1.1$$

## 158. d

50% of the parts will be below the mean diameter of 0.500 in.

0.504 in. is $2\sigma$ to the right of the mean. From *Fundamentals of Manufacturing*, 2nd Edition (Rufe 2002), Figure 43-4 (p. 322), 95.46% of the parts will fall between 0.500 in. $\pm 2\sigma$. Therefore, the percentage of parts between 0.500 in. and 0.504 in. is 0.5(95.46%) = 47.73%.

The percentage of parts smaller than 0.500 in. = 50%.

The percentage of parts smaller than 0.504 in. is 50% + 47.73% = 97.73%.

Out of 1,000 parts that equals 0.9773 × 1,000 parts ≈ 978 parts.

## 159. a

Based on *Fundamentals of Manufacturing*, 2nd Edition (Rufe 2002), Eq. 43-12 (p. 327), and from the choices given, *reducing variation* is the only way to increase process capability.

## 160. d

From *Fundamentals of Manufacturing*, 2nd Edition (Rufe 2002), Eq. 43-14 (pp. 327–328), a high $C_p$ means the process has low variability and a low $C_{pk}$ means the process distribution is skewed either left or right of the nominal.

## 161. c

From *Fundamentals of Manufacturing*, 2nd Edition (Rufe 2002), pp. 328–329, acceptance sampling plans are not required if suppliers ship zero defects.

## 162. a

From *Fundamentals of Manufacturing*, 2nd Edition (Rufe 2002), p. 337, *parallax error* refers to measurement error due to the position of the operator with respect to the instrument.

## 163. d

Based on *Fundamentals of Manufacturing*, 2nd Edition (Rufe 2002), Example 44.7.2, p. 339,

Gage tolerance = .1 × .020 = .002 in.
Wear allowance = .05 × .020 = .001 in.

The go side must fit over the largest shaft diameter within specification (1.010 in.). This diameter is decreased by the wear allowance, so the gage approaches the largest acceptable shaft diameter as it wears.

Go side = 1.010 − 0.001 = 1.009 in.

The gage tolerance is unilaterally applied so the gage will reject good parts rather than accept bad parts.

Go side = 1.009 in.$^{+.000 \text{ in.}}_{-.002 \text{ in.}}$

The no-go side should not fit over any shaft within specification. The no-go side is the smallest acceptable shaft diameter with the gage tolerance applied unilaterally.

No-go side = .990 in.$^{+.002 \text{ in.}}_{-.000 \text{ in.}}$

## Solutions

**164. a**

From *Fundamentals of Manufacturing*, 2nd Edition (Rufe 2002), pp. 84–88, a *tensile test* is a destructive test. Nondestructive testing is discussed in the *Fundamentals of Manufacturing Supplement* (Rufe 2005), pp. 124–128.

**165. a**

From *Fundamentals of Manufacturing*, 2nd Edition (Rufe 2002), p. 331, the probable reason is the measurement instrument lacks precision. It is unlikely that any process can produce 100 parts exactly the same. The instrument may be very accurate; however, it does not have enough precision to measure the differences between the shaft diameters.

**166. d**

From *Fundamentals of Manufacturing*, 2nd Edition (Rufe 2002), p. 338, qualification of internal diameters can be done with a *plug gage*.

**167. b**

From *Fundamentals of Manufacturing*, 2nd Edition (Rufe 2002), pp. 333–334, the diameter of the shaft measured by the 0–1.0-inch micrometer is 0.116 in.

**168. d**

From *Fundamentals of Manufacturing*, 2nd Edition (Rufe 2002), p. 334, *dial indicators* measure based on comparison as opposed to measuring directly.

**169. d**

From *Fundamentals of Manufacturing*, 2nd Edition (Rufe 2002), pp. 335–336, an *optical comparator* is a device for measuring small and/or complex parts. It projects a magnified image on a screen.

**170. b**

From *Machinery's Handbook*, 23rd edition, p. 713 (Oberg et al. 1988), the process of combining gage blocks is known as *wringing*. This answer can also be found on p. 12-9 of the *TMEH Desk Edition* (Cubberly and Bakerjian 1988).

## MANUFACTURING MANAGEMENT

### 171. b

From *Fundamentals of Manufacturing*, 2nd Edition (Rufe 2002), p. 350, and *Fundamentals of Manufacturing Supplement* (Rufe 2005), p. 141, in the *line-and-staff organization* channels of communication are well established.

### 172. c

From *Fundamentals of Manufacturing*, 2nd Edition (Rufe 2002), p. 350, *decentralization* allows more decisions to be made at lower levels.

### 173. d

From *Fundamentals of Manufacturing*, 2nd Edition (Rufe 2002), p. 352, *Pareto analysis* can determine key process variables to chart or monitor.

### 174. b

From *Fundamentals of Manufacturing*, 2nd Edition (Rufe 2002), p. 353, good teamwork is the result of many factors, including *positive reinforcement*.

### 175. a

From *Fundamentals of Manufacturing*, 2nd Edition (Rufe 2002), pp. 353–354, according to *participatory management*, employees are empowered to implement their decisions.

### 176. c

From *Fundamentals of Manufacturing Supplement* (Rufe 2005), p. 162, a company can acquire protection for a brand name through a *trademark*.

### 177. d

From *Fundamentals of Manufacturing*, 2nd Edition (Rufe 2002), p. 358, *collective bargaining* permits labor to strike after 60 days of notice to management.

### 178. b

From *Fundamentals of Manufacturing*, 2nd Edition (Rufe 2002), p. 358, a *union shop* requires employees to join the union after a specified period of employment.

### 179. c

From *Fundamentals of Manufacturing Supplement* (Rufe 2005), p. 173, type "C" fire extinguishers can be used on electrical equipment fires. This answer can also be found on p. 50-16 of the *TMEH Desk Edition* (Cubberly and Bakerjian 1988).

## 180. c

From *Fundamentals of Manufacturing*, 2nd Edition (Rufe 2002), p. 360, MSDS sheets must be readily available to all employees.

## 181. b

From *Fundamentals of Manufacturing*, 2nd Edition (Rufe 2002), p. 365, tenosynovitis is a repetitive motion injury.

## 182. b

From *Fundamentals of Manufacturing*, 2nd Edition (Rufe 2002), Table 46-1 (p. 363), based on women in the 5th percentile, standing assembly workers can be expected to reach 73.0 in. (1,854 mm).

## 183. d

From *Fundamentals of Manufacturing*, 2nd Edition (Rufe 2002), Table 46-2 (p. 364), the minimum recommended illumination for an office is 100 fc (1,076 lux) due to the need to read handwritten text.

## 184. c

From *Fundamentals of Manufacturing*, 2nd Edition (Rufe 2002), pp. 363–364, hearing protection is not necessary if noise abatement reduces the noise to an acceptable level.

## 185. d

From *Fundamentals of Manufacturing*, 2nd Edition (Rufe 2002), p. 359, on-the-job safety and health is achieved if integrated throughout the business or organization.

## 186. a

$$F = 3,000$$
$$n = 4$$
$$i = 12\%$$

Referencing *Fundamentals of Manufacturing*, 2nd Edition (Rufe 2002), Eq. 47-1 (p. 368),

$$P = F(P/F, i, n)$$

and Table 47-1 (p. 368),

$$(P/F, i, n) = \frac{1}{(1+i)^n}$$

$$P = 3,000 \left( \frac{1}{(1+.12)^4} \right) = \$1,907$$

The proportionality factor, $(P/F, i, n)$, also can be obtained from the *Fundamentals of Manufacturing Supplement* (Rufe 2005), Appendix E (p. 211).

## 187. c

$P = 5{,}000$
$A = 500$
$n = 5$
$F = 600$
$i = 8\%$

$EUAC$ = equivalent uniform annualized cost

$$EUAC = P(A/P,i,n) + A - F(A/F,i,n)$$

Referencing *Fundamentals of Manufacturing*, 2nd Edition (Rufe 2002), Eq. 47-1 and Table 47-1 (p. 368) or *Fundamentals of Manufacturing Supplement* (Rufe 2005), Appendix E (p. 207),

$$(A/P,i,n) = \frac{i(1+i)^n}{(1+i)^n - 1}$$

$$(A/F,i,n) = \frac{i}{(1+i)^n - 1}$$

$$EUAC = 5{,}000\left(\frac{0.08(1+0.08)^5}{(1+0.08)^5 - 1}\right) + 500 - 600\left(\frac{0.08}{(1+0.08)^5 - 1}\right)$$

$$= \$1{,}650$$

## 188. d

From *Fundamentals of Manufacturing*, 2nd Edition (Rufe 2002), p. 370, decreasing the parts in an assembly will decrease the variable cost.

## 189. c

Referencing *Fundamentals of Manufacturing Supplement* (Rufe 2005), Eq. 55-7 (p. 137),

$$\sum R = \sum C$$

$R$ = revenue
$C$ = cost

Eq. 55-8 (p. 137),
*Revenue*

$R = r \times Q \times n$

$r$ = \$30 per gear
$n$ = 1 month

$R = 30 \times Q \times 1 = 30Q$

and Eq. 55-9 (p. 137),
*Cost*

$V = v \times Q \times n$

$v = 10 + 5 = 15$ per gear
$n$ = 1 month

$V = 15 \times Q \times 1 = 15Q$

Overhead = \$25,000 per month, therefore

$30Q = 15Q + 25{,}000$
$15Q = 25{,}000$
$Q = 1{,}667$ parts

Minimum capacity needed =

$$\frac{1{,}667}{2{,}000} \times 100 = 83.4\%$$

**190. c**

From *Fundamentals of Manufacturing*, 2nd Edition (Rufe 2002), pp. 371–372, in *value engineering and analysis* teams work to achieve required functions without sacrificing manufacturability.

# PERSONAL/PROFESSIONAL EFFECTIVENESS

**191. c**

From *Fundamentals of Manufacturing Supplement* (Rufe 2005), pp. 1–2, a good listener must not focus on the delivery style.

**192. a**

From *Fundamentals of Manufacturing Supplement* (Rufe 2005), p. 5, when comparing data with respect to time, a line chart and/or bar chart should be used.

**193. d**

From *Fundamentals of Manufacturing Supplement* (Rufe 2005), p. 14, prior to a meeting the person facilitating it must determine the meeting's objectives.

**194. a**

From *Fundamentals of Manufacturing Supplement* (Rufe 2005), p. 12, a successful negotiation occurs when the interests of both sides are satisfied.

**195. c**

From *Fundamentals of Manufacturing Supplement* (Rufe 2005), p. 13, *confrontation* can relieve stress. The other choices provided are not always true.

**196. c**

From *Fundamentals of Manufacturing Supplement* (Rufe 2005), pp. 16–17, *creative thinking* can only be proven through implementation.

**197. a**

From *Fundamentals of Manufacturing Supplement* (Rufe 2005), p. 3, readers cannot infer the author's tone from his/her body language since it is written communication.

**198. a**

From *Fundamentals of Manufacturing Supplement* (Rufe 2005), pp. 15–16, the engineering supervisor should prioritize the topics and save the less urgent topics for the next meeting.

**199. c**

From *Fundamentals of Manufacturing Supplement* (Rufe 2005), p. 2, the best format for communicating a project summary to upper management would be a report.

**200. d**

From *Fundamentals of Manufacturing Supplement* (Rufe 2005), pp. 10–11, the style of communication will most likely be formal with no audience interaction.

## REFERENCES

Cubberly, William H. and Bakerjian, Ramon, eds. 1988. *TMEH Desk Edition.* Dearborn, MI: Society of Manufacturing Engineers.

Oberg, Erik, Jones, Franklin D., Horton, Holbrook L., and Ryffel, Henry H. 1988. *Machinery's Handbook*, 23rd Edition. New York: Industrial Press.

Rufe, Philip D., ed. 2002. *Fundamentals of Manufacturing*, 2nd Edition. Dearborn, MI: Society of Manufacturing Engineers.

Rufe, Philip D., ed. 2005. *Fundamentals of Manufacturing Supplement.* Dearborn, MI: Society of Manufacturing Engineers.